OXFORD HANDBOOKS IN E
Series Editors R. N. Illingworth, C.

2. Accidents and Emergencies in Children
   ROSEMARY J. MORTON AND BARBARA M. PHILLIPS
3. The Management of Wounds and Burns
   JIM WARDROPE AND JOHN A. R. SMITH
4. Cardiopulmonary Resuscitation
   DAVID V. SKINNER AND RICHARD VINCENT
5. The Management of Head Injuries
   DAVID G. CURRIE
6. Anaesthesia and Analgesia in Emergency Medicine
   KAREN A. ILLINGWORTH AND KAREN H. SIMPSON
7. Maxillofacial and Dental Emergencies
   JOHN E. HAWKESFORD AND JAMES G. BANKS
8. Emergencies in Obstetrics and Gynaecology
   LINDSEY STEVENS
9. The Management of Major Trauma (Second edition)
   COLIN ROBERTSON AND ANTHONY D. REDMOND
10. Environmental Medical Emergencies
    DAVID J. STEEDMAN

OXFORD HANDBOOKS IN EMERGENCY MEDICINE

This series has already established itself as the essential reference series for staff in A & E departments.

Each book begins with an introduction to the topic, including epidemiology where appropriate. The clinical presentation and the immediate practical management of common conditions are described in detail, enabling the casualty officer or nurse to deal with the problem on the spot. Where appropriate a specific course of action is recommended for each situation and alternatives discussed. Information is clearly laid out and easy to find—important for situations where swift action may be vital.

Details on when, how, and to whom to refer patients is covered, as well as the information required at referral, and what this information is used for. The management of the patient after referral to a specialist is also outlined.

The text of each book is supplemented with checklists, key points, clear diagrams illustrating practical procedures, and recommendations for further reading.

The Oxford Handbooks in Emergency Medicine are an invaluable resource for every member of the A & E team, written and edited by clinicians at the sharp end.

Environmental Medical Emergencies

David J. Steedman
*Consultant in Accident and Emergency
Medicine, Royal Infirmary of Edinburgh*

Oxford • New York • Tokyo
OXFORD UNIVERSITY PRESS
1994

Oxford University Press, Walton Street, Oxford OX2 6DP
Oxford   New York   Toronto
Delhi   Bombay   Calcutta   Madras   Karachi
Kuala Lumpur   Singapore   Hong Kong   Tokyo
Nairobi   Dar es Salaam   Cape Town
Melbourne   Auckland   Madrid
and associated companies in
Berlin   Ibadan

Oxford is a trade mark of Oxford University Press

Published in the United States
by Oxford University Press Inc., New York

© David J. Steedman, 1994

All rights reserved. No part of this publication may be
reproduced, stored in a retrieval system, or transmitted, in any
form or by any means, without the prior permission in writing of Oxford
University Press. Within the UK, exceptions are allowed in respect of any
fair dealing for the purpose of research or private study, or criticism or
review, as permitted under the Copyright, Designs and Patents Act, 1988, or
in the case of reprographic reproduction in accordance with the terms of
licences issued by the Copyright Licensing Agency. Enquiries concerning
reproduction outside those terms and in other countries should be sent to
the Rights Department, Oxford University Press, at the address above.

This book is sold subject to the condition that it shall not,
by way of trade or otherwise, be lent, re-sold, hired out, or otherwise
circulated without the publisher's prior consent in any form of binding
or cover other than that in which it is published and without a similar
condition including this condition being imposed
on the subsequent purchaser.

A catalogue record for this book is available from the British Library

Library of Congress Cataloging in Publication Data
Steedman, David J.
Environmental medical emergencies / David J. Steedman.–1st ed.
p.   cm.–(Oxford handbooks in emergency medicine ; 10)
Includes bibliographical references.
1. Medical emergencies–Handbooks, manuals, etc.
2. Environmentally induced diseases–Handbooks, manuals, etc.
3. Accidents–Handbooks, manuals, etc.   I. Title.   II. Series.
RC86.8.S73   1994   616.9'89025–dc20   94–7087
ISBN 0 19 262393 1 (hbk)
ISBN 0 19 262392 3 (pbk)

Typeset by Footnote Graphics, Warminster, Wiltshire
Printed in Great Britain on acid-free paper by
Biddles Ltd, Guildford and King's Lynn

Preface

Enthusiasm for outdoor pursuits and burgeoning interest in environmental issues has led to an increasing awareness of the potential for hazardous interaction between man and his surroundings. However, the successful management of environmental accidents and emergencies is often compromised by widespread misunderstanding, and procrastination will permit a relentless downward slide. Casualties from environmental injury require intervention that is both urgent and predictable on basic pathophysiological principles.

Some restriction has necessarily been placed on the definition of environment as this handbook has not been conceived as a comprehensive treatise. Selected topics have been chosen covering many of the common problems encountered by humans in various environments to which they are exposed.

This handbook is aimed at the 'man on the spot' who is hardly ever a specialist. It is not intended for the expert and no one will become an expert by reading it. Faced with the management of victims of environmental injury, doctors working in Accident and Emergency departments will find a source of practical help when a search through a more authoritative text is impracticable. It contains some compromises in an attempt to simplify as far as possible the management of clinical situations that are often far from simple. However, sources of assistance are also provided and each section includes a list of further reading.

The optimal management of environmental injury involves a continuum of care which commences with rescue and resuscitation on site. Each topic includes a section on management at the scene which will be of value to medical and paramedical personnel involved in prehospital care. Doctors from a variety of critical care disciplines will also find areas directly appropriate to their practice.

A succession of recent major accidents, both natural and man-made has brought the need for proper planning of emergency medical care into focus. The practical problems associated with casualties from selected environmental disasters are covered.

Reference to other handbooks in this series on cardio-pulmonary resuscitation, children's accidents and emergencies, and major trauma will complement the information presented in this handbook.

Finally, although preventive strategies are not covered in detail, public safety campaigns, community education, and observation of safety standards will reduce the number of environmental accidents and therefore lessen the need to resort to the management guidance in the text which follows.

Edinburgh D.J.S.
September 1993

Acknowledgements

I would like to thank the following specialists for their invaluable advice:

Evan Lloyd Consultant Anaesthetist Western General Hospital, Edinburgh	Hypothermia. Cold injury.
Mark Harries Consultant Physician Northwick Park Hospital, Middlesex	Near drowning.
Alan Milne Senior Medical Adviser British Antarctic Survey Medical Unit	Diving emergencies.
Iain Ledingham Dean, Faculty of Medicine and Health Sciences, United Arab Emirates University	Heat illness.
Charles Clarke Director, Mountain Medicine Centre St. Bartholomew's Hospital, London	High-altitude illness.
Derek Elsom Director, Research Centre of the Tornado and Storm Research Organisation	Lightning injuries.
Christopher Kalman Chief Medical Officer, Scottish Nuclear Ltd	Radiation accidents.
Alexander Proudfoot Director, Scottish Poisons Information Bureau	Poisoning by plants and fungi. Venomous bites and stings. Chemical accidents and emergencies.

I would also like to thank Robin Illingworth for his helpful criticism and the staff at Oxford University Press for their support.

Dose schedules are being continually revised and new side-effects recognized. Oxford University Press makes no representation, express or implied, that the drug dosages in this book are correct. For these reasons the reader is strongly urged to consult the drug company's printed instructions before administering any of the drugs recommended in this book.

Contents

1	Hypothermia	1
2	Cold injury	27
3	Near drowning	39
4	Diving emergencies	51
5	Heat illness	69
6	High-altitude illness	89
7	Lightning injury	105
8	Chemical accidents and emergencies	115
9	Radiation accidents	139
10	Poisoning by plants and fungi	165
11	Venomous bites and stings	187
	Appendix: Sources of help	201
	Index	207

CHAPTER 1

Hypothermia

General introduction	3
Regulation of body temperature	5
Clinical features	7
Problems after rescue	11
Rewarming techniques	13
At the scene	19
Initial hospital management	22
Further reading	25

Key points in hypothermia

1. The clinical features of mild hypothermia are frequently subtle and the diagnosis is often overlooked.
2. Hypothermia can mimic or mask other clinical disorders so the patient should be normothermic before a diagnosis is made.
3. Rapid rewarming can cause reversal of fluid shifts and precipitate pulmonary and cerebral oedema especially in the elderly.
4. Surface rewarming can cause cardiovascular collapse if there is inadequate circulating blood volume.
5. Airway warming is an effective non-invasive method which can be used in combination with other techniques.
6. Pulses may be difficult to detect in profoundly hypothermic patients so careful palpation is necessary before commencing external chest compressions.
7. Ventricular fibrillation may be precipitated in victims with profound hypothermia by rough handling or inappropriate therapeutic intervention.
8. Death from hypothermia can only be diagnosed with certainty when a victim fails to respond to resuscitation after rewarming.

General introduction

- **Definition of hypothermia   Immersion hypothermia
  Exhaustion hypothermia   Urban hypothermia**

Accidental hypothermia can contribute to death, illness or injury in a wide range of environmental situations. Primary accidental hypothermia occurs despite normal thermoregulation on exposure to cold which is overwhelming. Secondary accidental hypothermia develops in mild and moderate cold exposure because of abnormal thermogenesis. Hypothermia induced deliberately as part of a therapeutic regime (e.g. cardiac surgery) will not be considered here.

Cold stress applies to any degree of environmental cold which causes the physiological thermoregulatory mechanisms to be activated. The body aims to maintain the temperature of the vital organs in the inner core as constant as possible and generally succeeds over a wide range of thermal stresses. The body core temperature is preserved at the expense of the outer shell which consists of the superficial tissues and usually makes up about 10 per cent of the body mass.

Hypothermia occurs when the temperature of the body core falls below 35 °C

There are three groups of hypothermic patients commonly encountered by the emergency services. The physiological changes are different in each group and so the clinical management varies accordingly.

Immersion hypothermia develops rapidly when the cold stress is so great that the core temperature is forced down despite maximal heat production. Hypothermia therefore occurs before the body becomes exhausted. The most common cause of this type of hypothermia is falling into cold water. The average sea temperature in Britain is 11 °C during summer. Cases of drowning and near drowning are often the consequences of the hypothermic victim losing consciousness. Scuba divers are usually in negative heat balance despite wearing a wet or dry suit for thermal protection. Hypothermia is involved in more than 20 per cent of scuba

4 • Hypothermia

diving fatalities. Accidental hypothermia in cavers often involves immersion in cold water.

Exhaustion hypothermia develops when the body's fuel stores become depleted. The cold stress is such that the heat production can maintain core temperature provided sufficient energy stores are available. This type of hypothermia is most commonly found in mountaineering and hill walking accidents. Fifteen per cent of surviving casualties rescued from Scottish mountains between 1967 and 1977 were hypothermic. Although only 10 per cent of deaths were attributed to hypothermia many of the 75 per cent of people recorded as dying from physical injuries probably suffered from exposure to cold which may have contributed to the fatal outcome.

More widespread participation in outdoor winter sports has also resulted in an increased number of patients with exhaustion hypothermia. An unforeseen event such as a minor injury or equipment failure can overtake the victim whilst engaged in such activities in potentially dangerous cold surroundings. The runner, training or competing in long distance events during winter, is also susceptible. The high rate of heat production will offset the massive heat loss which occurs while running. However, decreased energy reserves, exhaustion, dehydration, and evaporation of sweat can precipitate rapid onset of hypothermia after stopping.

Urban hypothermia is commonly seen in the elderly following prolonged exposure to cold of a mild degree. The core temperature may remain above 35°C for several weeks. However, heat production is eventually insufficient to counteract the cold and hypothermia supervenes. Associated factors such as a fall may precipitate the development of hypothermia.

Altered behavioural patterns in people taking drugs or alcohol can be a precipitating factor in all types of hypothermia.

Regulation of body temperature

- Routes of heat loss   Central regulation   Heat production   Regulation of heat loss

Routes of heat loss

The extent of heat loss which occurs through the normal routes (Box 1.1) is affected by a number of factors.

Heat loss via infrared *radiation* is maximal when the body is unclothed and least when curled up and insulated.

Conduction is heat exchange which occurs during direct contact. The amount of heat transferred by conduction depends on the thermal gradient between the two surfaces. Conductive heat losses may increase 25-fold in water compared to air.

Box 1.1   Routes of heat loss

- Radiation
- Conduction
- Convection
- Evaporation

Convection occurs when air or water is heated by being in close proximity to the body and is then removed by currents. Convective heat loss is increased by limb movement and shivering.

Evaporative losses develop during conversion of a liquid to its gaseous phase. Loss occurs from the skin whilst sweating and from the respiratory tract during warming and humidification of inspired air.

Both convective and evaporative losses are increased considerably in windy conditions which has resulted in the use of the term 'wind-chill' (see Chapter 2, pp. 29–31). For example, a 16 km per hour wind will double the rate of heat loss at $-1\,°C$ compared to still conditions.

Central regulation

Core temperature is controlled by a central thermostat situ-

6 • Hypothermia

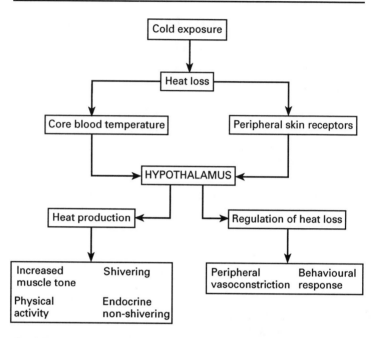

Fig. 1.1 • Physiology of cold exposure and temperature regulation

ated in the hypothalamus (Fig. 1.1) and is normally regulated within a narrow range. The thermostat is activated by impulses from central receptors which respond to changes in blood temperature and also from peripheral skin receptors.

The hypothalamus regulates the core temperature by governing heat production and heat loss.

Heat production

Basal heat production generates about 100 kilocalories per hour in an averaged sized adult. Increase in muscle tone can double this heat production. Maximal heat production generated by shivering can increase heat production to five times the basal rate. Glycogen depletion and fatigue limit the duration of maximal heat production from shivering to a few hours. Aerobic fitness may permit free fatty acids to be utilized as an alternative source of energy and prolong maximal shivering thermogenesis. Shivering is suppressed by hypoglycaemia, hypoxia, fatigue, phenothiazines, barbiturates, and alcohol.

Physical activity will also increase heat production up to 10- to 15-fold over basal rates during maximal exercise. However, such vigorous activity may be restricted by injury, severe cold stress, exhaustion, malnutrition, and at high altitude, by hypoxia.

Only 50 per cent of the extra heat generated during shivering and deliberate physical activity is retained in the body because there is associated heat loss from increased muscle blood flow.

Heat may also be produced without shivering or muscle activity. Metabolic activities controlled by catecholamines and endocrine hormones appear to play a role in this so-called non-shivering thermogenesis.

Regulation of heat loss

Following exposure to cold stress the body responds by peripheral vasoconstriction mainly via the sympathetic nervous system. Heat loss is reduced as blood is diverted away from the skin surface to the core. Heat loss from the head may, however, be considerable as there is minimal vasoconstrictor activity at this site. The head is therefore also relatively resistant to frostbite compared to the limbs.

Heat loss may be decreased by behavioural responses, e.g. clothing, getting out of the wind, and huddling together in a group. Curling up into the fetal position reduces the total surface area exposed to the cold environment by 30–40 per cent and areas of particularly high heat loss are protected.

Clinical features

- **Mild and profound hypothermia   Differentiation from death**

Mild and profound hypothermia

Hypothermia can be divided into *mild* (core temperature

8 • Hypothermia

32°C or above) and *profound* (core temperature less than 32°C). The management and prognosis differ markedly between these two groups.

The signs and symptoms of mild hypothermia reflect the body's normal physiological responses to maintain core temperature. In profound hypothermia, clinical features result from the pathological changes which occur as the body systems are cooled. Increasing disturbance of cerebral and cardiac function appear as the temperatures of the heart and brain fall below 35°C (Table 1.1). Individuals vary widely in their responses to hypothermia. For example, consciousness is usually lost at about 30°C but sometimes coma does not occur until the core temperature is as low as 26°C. Hypothermia is unlikely to be the cause of coma if the core tempera-

Table 1.1 • Clinical features of hypothermia

	Temperature (°C)	
	37.6	'Normal' rectal temperature
	37	'Normal' oral temperature
	36	Basal metabolic rate increases
	35	Maximal shivering
Mild	34	Amnesia, dysarthria, ataxia
		Blood pressure maintained
	33	Retrograde amnesia, dilated pupils
Profound	<32	Conscious level progressively depressed
	31	Hypotension
	30	Increased muscle rigidity
	29	Bradycardia, respiratory depression
	28	Ventricular fibrillation may develop if heart irritated
	27	Voluntary motion lost along with pupillary light and deep tendon reflexes
	25	Ventricular fibrillation may develop spontaneously
	23	No corneal reflexes
	20	Asystole
	19	Isoelectric electroencephalogram
	18	Lowest recorded temperature in an accidental hypothermic patient with recovery.

ture is above 33 °C. Similarly, shivering has been recorded down to a core temperature of 24 °C and some patients may develop profound hypothermia without shivering.

The signs and symptoms of mild hypothermia are frequently subtle and may be overlooked. Hypothermia may mimic drug or alcohol overdose, cerebrovascular accidents, head injury, post-ictal states, and hypoglycaemic coma. These conditions may also present in association with hypothermia when the clinical features may be masked.

The bradycardia which occurs during hypothermia is due to decreased depolarization of pacemaker cells and is resistant to atropine. Shivering may produce a baseline artefact and obscure P-waves. Below 32 °C, both atrial and ventricular arrhythmias develop from re-entrant pathways and ectopic foci. Prolongation of the PR interval, the QRS complex, and characteristically the QT interval, reflects the hypothermic depression of the cardiac conducting tissue. J-waves which are potentially diagnostic may appear at temperatures below 32 °C and are seen at the junction of the QRS complex and ST segments. They are probably due to hypothermic induced ion fluxes causing delayed LV depolarization or early repolarization (Fig. 1.2).

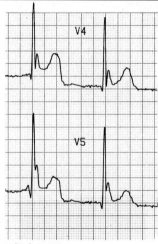

Fig. 1.2 • Hypothermic J-waves

10 • Hypothermia

Although hypothermia protects the brain from the effects of hypoxia, in clinical practice survival is almost totally dependent on maintaining sufficient blood pressure to perfuse the heart and brain. Cardiac function is therefore more relevant to ultimate survival than cerebral temperature.

Metabolism in the liver slows during hypothermia and lactate and other products of muscle metabolism accumulate contributing to a metabolic acidosis. Detoxification and excretion of drugs may also be considerably reduced so that drugs normally metabolized in this way must be given in smaller doses.

A rise in serum amylase is common in hypothermia but clinical evidence of pancreatitis is only occasionally seen after severe hypothermia.

The blood glucose level may be high or low in hypothermic patients. Increase in the rate of breakdown of hepatic glucose occurs during the initial stages of acute exposure due to catecholamine release. Below 30°C, insulin release and activity are reduced and target cells are insulin-resistant. The hyperglycaemia that results resolves spontaneously on rewarming. Persistence of hyperglycaemia following rewarming may signify pancreatitis. Chronic exposure following exhaustion and glycogen depletion leads to hypoglycaemia.

Thrombocytopenia and neutropenia occur in hypothermia and can be pronounced below 28°C. The low platelet and neutrophil count, which is due to sequestration in the liver, spleen, and splanchnic bed, usually returns to normal on rewarming. In severe hypothermia disseminated intravascular coagulation may develop.

Differentiation from death

Profound hypothermia may present with a clinical picture which is very difficult to distinguish from death; the skin may be ice cold, muscles and joints in a state of apparent rigor mortis, respiration and major pulses are difficult to detect, pupillary and other reflexes absent, and the patient is unresponsive. The electrocardiograph may show a broad complex bradyarrhythmia. The total cessation of cardiorespiratory activity is compatible with survival and a flat electroencephalogram is not a certain indicator of death. *The only certain diagnosis of death in hypothermia is failure to recover on rewarming.*

Problems after rescue

- Core temperature afterdrop   Precipitation of ventricular fibrillation   Fluid volume changes

Core temperature afterdrop

After a victim has been removed from a cold stress the core temperature may continue to drop before starting to rise again. This phenomenon is known as the afterdrop. Two processes contribute to this afterdrop in core temperature.

The first is a simple process of temperature equilibration between the core and the periphery which continues until a lower average temperature is attained. The second mechanism is the counter-current cooling of blood which occurs when blood perfuses cold tissues. The core temperature will continue to fall until the temperature gradient is eliminated. Heat applied to the skin surface causing perpheral vasodilation, or exercise causing an increase in the circulation through cold muscle can both exacerbate this process.

There is no evidence to support the theory that the afterdrop is caused by a cold bolus of blood returning to the core from the cooler peripheral circulation. There is no peripheral pool of blood because of the intense peripheral vasoconstriction.

Precipitation of ventricular fibrillation

Cardiac irritability and associated risk of ventricular fibrillation become greater as the cardiac temperature is lowered. An afterdrop in temperature will therefore increase the likelihood of arrhythmias. There are a number of factors which may trigger the onset of ventricular fibrillation. Sudden changes in hydrogen and electrolyte concentrations and $PaCO_2$, as well as hypoxia are implicated.

During rewarming from hypothermia any factor which alters the intracardiac thermal gradient so that the subendocardial and neuromuscular conducting system become

12 • Hypothermia

significantly different from the main mass of cardiac muscle increases the risk of ventricular fibrillation.

Mechanical irritation is also liable to precipitate ventricular fibrillation. This may develop if the victim is roughly handled during evacuation or when being lifted or rolled on to a stretcher or trolley.

Difficulty may be encountered in palpating a major pulse in a hypothermic victim especially if there is a bradyarrhythmia. Commencing external chest compressions in such patients may similarly irritate the myocardium and precipitate arrhythmias.

Fluid volume changes

In response to a cold stress, impulses from the skin and central receptors reinforce reflex peripheral vasoconstriction. When the cold stress is removed relaxation of the vasoconstrictor tone lowers vascular resistance and increases the volume of the vascular bed. If the central circulating blood volume is insufficient to meet this increase, central venous and arterial blood pressure will fall. Cardiovascular collapse is well documented during rewarming of hypothermic victims.

Cold induced vasoconstriction shunts blood into the deep capacitance vessels and the relative excess volume is removed by a diuresis. The dehydration which results may be exacerbated further by exercise which increases intravascular volume and maintains a diuresis.

Box 1.2 Problems following rescue

- Core temperature afterdrop
- Precipitation of ventricular fibrillation
- Hypovolaemic collapse
- Compartmental fluid shifts
- Release of hydrostatic squeeze

During cooling there is a shift in compartmental body fluids from the intravascular space into the cells and interstitium. The severity of the shifts is directly related to the duration of cold exposure. During acute immersion hypothermia there is insufficient time for a major shift to develop and therefore hypotension rarely occurs during rewarming. On the other hand, in exhaustion hypothermia, fluid shifts which develop over a longer time-scale, combined with dehydration from cold diuresis and losses from sweating and the respiratory tract, produce large volume deficits. During mountain rescue this may manifest clinically if the casualty is brought down in the 'head-up' position, producing significant orthostatic hypotension and loss of consciousness.

During rewarming the compartmental fluid shifts will reverse. Superficial tissues which were exposed to the cold stress for longer experience the greater fluid shifts both during cooling and rewarming. In cases of profound hypothermia treated only by exposure in a warm room, hypovolaemic collapse occurs because warm air provides a sensation of warmth with minimal heat gain to reverse the fluid shifts. Similarly, when external rewarming is achieved by a hot bath reversal of the vasoconstriction occurs. However, the intravascular expansion is compensated for by a reversal of the fluid shifts which occurs if the tissues are sufficiently rewarmed by this technique.

In rewarming elderly victims of urban hypothermia following prolonged exposure to cold, rapid reversal of fluid shifts can precipitate pulmonary and cerebral oedema. Appropriate monitoring is required in this group to prevent circulatory overload.

During rescue from cold water the hydrostatic squeeze is released as the victim is extricated. The venous return may suddenly fall and cause significant hypotension and cardiac arrest especially as baroceptor responses are impaired in hypothermia. A vertical rescue lift could result in further decrease in venous return due to gravitational effects.

Rewarming techniques

- **Spontaneous rewarming   Active rewarming**

The rewarming methods which are currently available can be used separately or in combination. The methods chosen

14 • Hypothermia

to rewarm an hypothermic victim will depend on the following:

- Degree of hypothermia
- Available equipment
- Clinical environment
- Clinical skills

Spontaneous rewarming

Spontaneous rewarming is achieved by preventing further heat loss while allowing the body to rewarm through endogenously generated metabolic heat. Covering the patient with an appropriate material stops evaporative and convective losses. The patient must be able to generate sufficient heat by shivering to maintain an adequate rate of rewarming. Shivering thermogenesis usually ceases below a core temperature of 32°C. When used in a hospital environment the ambient temperature should exceed 21°C. The technique is non-invasive and adequate for most patients with mild hypothermia (>32°C).

The 'space blanket' made of metallized plastic sheeting has no advantage over a similar thickness of polythene sheeting.

This technique maintains peripheral vasoconstriction reducing the incidence of vascular collapse due to rewarming. Rewarming rates of 0.5–2°C per hour can be achieved. In prolonged hypothermia in the elderly the rate should not exceed 0.5–0.6°C as too rapid a rate of spontaneous rewarming may precipitate rapid reversal of intercompartmental fluid shifts leading to cerebral and pulmonary oedema.

Box 1.3 **Methods of rewarming**

- Spontaneous rewarming
- Active rewarming — Surface
　　　　　　　　　　— Central
　　　　　　　　　　— Extracorporeal

Active rewarming

Active rewarming is indicated when endogenous thermogenesis is insufficient to produce an adequate rate of rewarming following insulation. Patients in cardiorespiratory arrest require rapid elevation of core temperature using active rewarming techniques. Patients with peripheral vasodilation induced by drugs or spinal cord injury should be actively rewarmed.

Surface heating

Surface heating using a hot bath with the water maintained between 37°C and 41°C is the most rapid method of active surface rewarming. This method has been advocated for use in victims of acute immersion hypothermia. Core temperature afterdrop associated with sudden peripheral vasodilation is a potential complication of this technique. Keeping the limbs out of the bath may allow some maintenance of peripheral vasoconstriction and minimize cardiovascular collapse. Although a hot bath can be used in victims with depressed consciousness, difficulty may be encountered in protecting the airway. Cardiopulmonary resuscitation is impossible and monitoring of the patient is compromised. The technique cannot be used in patients with external injuries.

Other methods of surface heating include plumbed garments which circulate warm fluids, hot water bottles, heating pads, and blankets applied to the trunk. A radiant heat cradle over the torso (with the skin protected by a thin sheet) can also be used.

Rewarming methods which raise skin temperature will depress shivering. Quantities of heat supplied by external rewarming may be insufficient to compensate for this. Treatment of profoundly hypothermic victims by exposure in a warm room without any other intervention is therefore hazardous. The sensation of warmth to the skin will also abolish vasoconstriction, resulting in a drop in blood pressure.

Central rewarming

Active core rewarming techniques decrease the probability

16 • Hypothermia

of cardiovascular collapse in patients presenting with core temperatures below 30°C.

Airway warming (respiratory insulation). The respiratory tract is an important site for heat exchange. Inhalation of heated humidified oxygen or air is an effective method of rewarming and prevents the loss of moisture and heat through breathing. Some additional heat is liberated as the water vapour condenses in the lungs. Dry air has a very low thermal conductivity and therefore humidification of the inspired gas is necessary. Inhalant temperatures of 40–45°C and an adequate respiratory minute volume are required for optimal heat delivery.

A simple and effective method of achieving airway warming utilizes the exothermic reaction between carbon dioxide and soda lime. If a person breathes through a container of soda lime the inspired air will be heated and humidified. The production of carbon dioxide is, however, decreased in a hypothermic victim and is insufficient to raise the temperature of the soda lime to an adequate level.

A simple circuit for providing airway warming is shown in Fig. 1.3. The patient may be connected by face mask or endotracheal tube as appropriate and ventilation assisted if required. The initial flow rate of carbon dioxide is 3–5 litres per minute with oxygen at 0.5–1 litre per minute. Once the heat in the Waters canister has reached a desired level, the carbon dioxide is discontinued. An initial charge of carbon dioxide should maintain the desired temperature

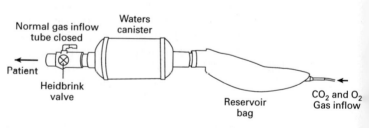

Fig. 1.3 • Circuit for airway warming

for approximately 30–60 minutes and additional carbon dioxide can be added at intervals if necessary. Care must be taken to monitor the inspired gas temperature which should be maintained at less than equal to 45°C to avoid thermal airway burns.

This system can be used with an open or non-return circuit in which all exhaled gases are vented to the atmosphere, or with a closed circuit which recycles the exhaled gases.

Airway warming should be used in conjunction with surface insulation and can be combined with other methods of core rewarming.

Airway warming is indicated in immersion and exhaustion hypothermic victims when the core temperature is 32°C or less. In elderly patients with 'subchronic' hypothermia airway rewarming should not be used without intensive therapy unit facilities because the accelerated rate of warming may precipitate cerebral and pulmonary oedema.

Airway warming is non-invasive and rewarming rates of 1–2.5°C per hour can be attained with minimal temperature afterdrop. This method has the additional advantage of ensuring adequate oxygenation and probably reduces the risk of arrhythmias by adequately warming the heart and hence the conducting system.

Peritoneal dialysis does not require expensive equipment and can be set up in a few minutes. Normal saline heated at 45°C is run into the peritoneal cavity through an infusion set. Heated fluid is left in the peritoneal cavity until most of its surplus heat has been given up to the tissues and it is then removed and replaced by a fresh warm supply.

Heat is conducted directly to the retroperitoneal structures including blood returning to the heart via the inferior vena cava. Hepatic rewarming re-activates depressed detoxification and conversion enzymes. Conductive heat transfer through the diaphragm provides additional heat to the heart and lungs. The method is contraindicated following intra-abdominal trauma.

Intravenous fluids and blood should be heated during

all hypothermic resuscitations. Efficient fluid warming systems are available. A Level 1 fluid warmer warms cold crystalloid solutions and blood from 10°C to 35°C via a heat exchange system at flow rates of 250–500 ml/min.

If the fluid warmer is not available intravenous fluids can be heated to 40–42°C in a microwave oven. This needs very careful control to avoid over-heating. Heating time should be determined for each individual microwave oven. The fluid should be mixed thoroughly prior to administration. Central administration of heated fluids can produce myocardial thermal gradients and precipitate arrhythmias. Thermal packs can be used in the field to ensure that cold fluid is not given.

Risk of fluid overload and precipitating pulmonary oedema limit the use of this method as a means of rewarming in pure hypothermia. For example, infusing 1 litre of fluid at 42°C into a 70 kg man will increase the core temperature of 28°C by 0.3°C.

Extracorporeal warming
The main advantage of rewarming by cardiopulmonary bypass is the preservation of blood flow to the brain and other vital organs in the presence of hypothermic cardiac arrest. However, cardiopulmonary bypass can only be performed by an experienced team and when the equipment is available. Consequently, cardiopulmonary bypass in practice is usually instituted after other active rewarming methods have been commenced. When the warmer is set at 38–40°C, femoral arterial flow rates of 2–3 litres/minute can elevate core temperature by 1–2°C every 3–5 minutes.

Potential complications of rapid rewarming in severe hypothermia include disseminated intravascular coagulation, pulmonary oedema, haemolysis, and acute tubular necrosis.

At the scene

- Diagnosis   Insulation   Rewarming   Resuscitation

Diagnosis

Death from hypothermia can only be confirmed with certainty when a victim fails to respond to cardiopulmonary resuscitation after rewarming. Death must never be diagnosed in the field although hazardous conditions may render prolonged attempts at resuscitation or evacuation to hospital impossible.

A history of the circumstances of discovery, duration of exposure, and past medical history are important in selecting optimal treatment.

Accurate field measurement of core temperature is usually not practicable and some thermometers are unreliable outdoors in cold ambient temperatures.

Insulation

Insulation of a patient from the environment to prevent further heat loss and to allow spontaneous rewarming can be provided for most hypothermic victims in the field by applying layers of polythene. A thick fibre-pile 'casualty bag' covered by a waterproof layer is even better. It is important when applying insulation to remember the head because up to 70 per cent of total heat production can be lost from this site depending on ambient temperature. Conduction losses will continue if a patient is simply covered with a blanket and no insulation provided between the body and the ground.

Wet clothing should be removed when the victim is in a warm shelter and out of the wind. If shelter is not available adding layers of dry clothing, especially a layer that is impervious to wind and water is recommended.

Rewarming

Rubbing the skin of the victim is absolutely contraindicated because it suppresses shivering by providing a sensation of warmth and increasing skin blood flow. No extra heat is added thus aggravating the cooling of the core.

Oral heated fluids are comforting but have no effect on raising body temperature. Alcohol suppresses shivering thermogenesis and should not be given.

Surface heating can be hazardous when performed in the field with inadequate monitoring. Surface rewarming suppresses shivering which impedes the rate of core rewarming.

Active rewarming in the field should probably be avoided unless evacuation is delayed or prolonged. Probably the only safe and practicable technique is airway warming (respiratory insulation). The Lloyd portable airway warming unit weighs 3 kg and consists of an oxygen cylinder, demand valve, and a 2 litre reservoir bag, and generates heat and moisture from a carbon dioxide and soda lime reaction. An inline thermometer measures average temperature of the inhaled gases.

Active surface rewarming may be the only option if a victim is isolated from medical care. Warm objects including commercially available hot packs may be placed against the groins, axilla, neck, or trunk although care should be taken to avoid burns. One method of warming used by mountain rescue teams is body to body contact inside a sleeping bag. If facilities are available, conscious shivering and uninjured casualties may be immersed in a hot bath. The temperature of the water should be comfortable to the elbow and the casualty and the temperature maintained by circulating additional hot water as necessary. Cessation of shivering will occur almost immediately but this should not be interpreted as an indication for removing the casualty.

Resuscitation

Oxygen should be given to hypothermic patients if available. Care should be taken not to leave the oxygen cylinder lying on cold ground or in snow. Changing a patient from warm expired air resuscitation to assisted ventilation with oxygen from a cold cylinder can precipitate ventricular fibrillation.

In profound hypothermia there is significant respiratory depression and the patient may appear apnoeic. If the patient is found in respiratory arrest, assisted ventilation should be commenced. Hyperventilation in hypothermic victims may induce hypocapnia and ventricular irritability. *Endotracheal intubation should not be undertaken unless spontaneous or assisted bag mask ventilation is inadequate and the protective reflexes are absent.* If necessary, pre-oxygenation may minimize the risk of precipitating ventricular fibrillation.

Palpation of peripheral pulses is difficult in hypothermic vasoconstricted patients when the cardiac rate is extremely slow. Careful palpation of the carotid or femoral pulse for at least one minute may be necessary. The low cardiac output may be sufficient to meet the metabolic demands in these patients and ventricular fibrillation can be precipitated by unnecessary external chest compressions. In hypothermic victims, chest compression should be performed at the same rate as in normothermic patients despite changes in the compressibility of the heart and compliance of the chest wall. If a monitor is available the cardiac rhythm should be determined. DC countershock at 2 J/kg should be given if indicated. However, attempts at defibrillation are usually unsuccessful until the core temperature is above 28–30 °C.

Intravenous access may be extremely difficult to achieve in the field because of the intense peripheral vasoconstriction.

Most drugs are ineffective in hypothermic patients or are dangerous to use because of increased cardiac irritability. Hypothermia can be precipitated by hypoglycaemia and hypoglycaemia can closely mimic the signs of hypothermia: 50 per cent dextrose should therefore be considered. Glucagon will be ineffective in glycogen-depleted patients. Correction of hypoglycaemia will only improve the level of consciousness to that expected for the current core temperature.

Carry the patient flat or head-down, not foot-down.

Prolonged field treatment should be avoided.

Initial hospital management

- Measurement of core temperature   Resuscitation
  Investigations

Measurement of core temperature

Adequate equipment should be available in the Accident and Emergency department to measure core temperature accurately. The normal clinical thermometer does not record lower than 35 °C and therefore cannot be used to diagnose hypothermia. Electronic thermometers using thermistors or thermocouple probes are more convenient to use than low reading mercury-in-glass thermometers. Ideally, measurement of core temperature should be made from the lower oesophagus but the rectum is often used as an alternative site. The rectal temperature may lag behind changes in core temperature and not reflect cardiac or brain temperature. Tympanic temperature which can be measured using infrared devices closely correlates with oesophageal temperature.

All clothing should be gently removed and cut off with minimal movement of the patient. Insulation against further heat loss can be achieved with layers of blankets and polythene. Warm the room and keep the door closed. Full examination and X-rays can wait until the patient is warm.

Assessment and management of the airway and ventilation follows the same principles as described in the pre-hospital setting.

Resuscitation

All hypothermic patients should have an intravenous cannula inserted. Most fluid shifts reverse during rewarming from mild hypothermia and intravenous infusion is not required. If the blood pressure falls during rewarming, a volume challenge of 300–500 ml of warmed normal saline should be administered. Further volume may be required depending on the response. Ringer's lactate solution should be avoided as the hepatic metabolism of lactate is depressed in hypothermia. Colloid solutions may be given if there is no response to crystalloid. Central venous pressure monitoring can be of value in unstable patients. However, catheter insertion may precipitate ventricular fibrillation. Fluid resuscitation is rarely required during recovery from acute immersion hypothermia.

One approach to rewarming patients with hypothermia is outlined in Fig. 1.4.

Investigations

Blood samples should be taken for glucose reagent strip estimation, laboratory blood glucose, full blood count, urea and electrolytes, serum amylase, and in profound cases of hypothermia arterial blood gas determinations and a coagulation screen are necessary. Toxicological investigation may be required depending on the circumstances of discovery.

Arterial blood samples are warmed to 37°C before electrode measurements are made. The pH can be corrected by adding 0.015 pH units per °C below 37°C. During correction, the PaO_2 drops 7.2 per cent per °C and the $PaCO_2$ drops 4.4 per cent per °C. However, the correction of pH does not take into account the serum protein buffering system which is less temperature-dependent than the carbonic acid–bicarbonate system. Clinical management should therefore be based on arterial blood gas measurements recorded at 37°C which is the best approximation of true blood pH at any body temperature.

Blood glucose is often high but falls with rewarming.

The cardiac rhythm should be monitored and a 12 lead electrocardiogram performed.

A chest X-ray is necessary to exclude aspiration and to monitor the development of pulmonary oedema during rewarming. If there is a history of blunt trauma, radiological investigation of the cervical spine is required.

Gastric dilation and poor gastric mobility are common in severe hypothermia and therefore a nasogastric tube should be passed. An indwelling urethral catheter is required in most patients with hypothermic collapse.

Ideally, all hypothermic patients should be transferred to an intensive therapy unit after initial management in the Accident and Emergency department.

24 • Hypothermia

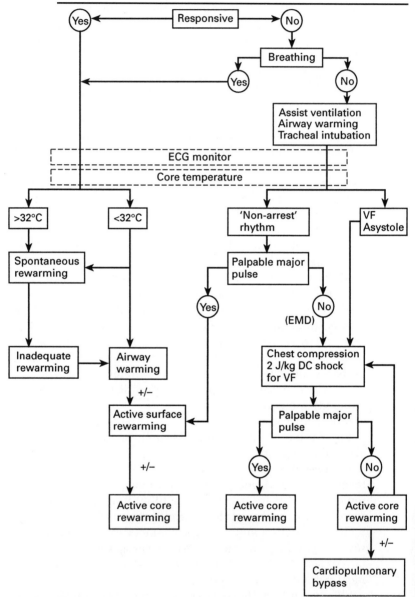

Fig. 1.4 • A schematic approach to the assessment and management of hypothermia

Further reading

Anonymous (1978). Treating accidental hypothermia. *British Medical Journal*, **ii**, 1383–4.

Bangs, C. C. (1984). Hypothermia and frostbite. *Emergency Medicine Clinics of North America*, **2**, 475–87.

Danzl, D. F. (1988). Accidental hypothermia. In *Emergency medicine, concepts and clinical practice*, (ed. P. Rosen, F. J. Baker, G. R. Braen, R. H. Dailey, and R. C. Levy), Vol. 1, pp. 663–92. Mosby, Boston.

Keatinge, W. R. (1991). Hypothermia: dead or alive. *British Medical Journal*, **302**, 3–4.

Lloyd, E. L. (1986). *Hypothermia and cold stress*. Croom Helm, London.

Lloyd, E. L. (1991). The management of accidential hypothermia. *Care of the Critically Ill*, **7**, 194–9.

Lloyd, E. L. (1991). Equipment for airway warming in the treatment of accidental hypothermia. *Journal of Wilderness Medicine*, **2**, 330–50.

Maclean, D. and Emslie-Smith, D. (1977). *Accidental hypothermia*. Blackwell Scientific, Oxford.

Medical Commission on Accident Prevention (1992). *Report of the working party on out of hospital management of hypothermia*. MCAP, London.

CHAPTER 2

Cold injury

General introduction	29
Non-freezing cold injury	30
Pathophysiology of frostbite	33
Clinical features of frostbite	34
At the scene	34
Immediate treatment in the Accident and Emergency department	35
Further reading	36

Key points in cold injury

1. Frostbite is caused by exposure to a freeze stress whereas chilblain and trench foot are produced by cold ambient temperatures above freezing.
2. Generally, frostbite develops over a period of hours resulting in extracellular ice crystal formation. Cell survival is therefore possible. Cellular crystal formation and immediate cell death only occur in very rapid freezing.
3. There should be no attempt to rewarm a frostbitten extremity in the field. If re-freezing occurs after thawing more tissue is ultimately lost.
4. Treatment of systemic hypothermia has priority over management of frostbite.
5. Rapid rewarming of the frozen area in a water bath 37–41°C is the treatment of choice.
6. Surgical debridement and excision should be delayed until spontaneous amputation of the soft tissues is complete which may take several months. Earlier surgical intervention may result in infection and greater tissue loss.

General introduction

- **Freezing vs. non-freezing cold injury   Risk factors**

Peripheral cold injury primarily affecting the extremities can occur in addition to the systemic effects of a cold stress which cause hypothermia. These injuries include frostbite, chilblain, and immersion or trench foot. Frostbite is caused by exposure to a freezing stress whereas chilblain and trench foot are provoked by cold temperatures above freezing.

Although cold injury occurs more frequently during military campaigns it is still a problem in civilian life. Cold injury is likely to increase in frequency as a result of greater participation in winter sports, mountaineering, and wilderness adventure.

Although ambient temperature and duration of exposure are key determinants of cold injury, other factors can influence the ultimate severity (Box 2.1).

Box 2.1 Determining factors in cold injury

Environmental factors — Ambient temperature
— Duration of exposure
— Wind-chill
— Wetness of skin

Host factors — Protective clothing
— Acclimatization
— Previous cold injury
— Local circulation
— Fatigue
— Impaired judgement
— Debility
— Injury

The influence of wind speed in increasing convective heat losses is well established and the 'wind-chill' index is a useful predictor of the risk of cold injury to unprotected areas. The wind-chill index corresponds to the degree of discomfort experienced. The effect can be quantified by quoting the equivalent still air temperature produced by any given wind speed (Table 2.1).

Contact of the skin with metal greatly increases conductive losses, especially if the hands are wet.

Cold injury rarely develops in healthy individuals with adequate and properly fitting protective clothing. Frostbite is almost always associated with a precipitating factor such as fatigue, immersion, immobilization, or accidental injury. These risks are increased when judgement is impaired and in one study alcohol was implicated in 50 per cent of victims.

Acclimatization to cold is protective against frostbite whereas previous cold injury increases the risk following subsequent exposure.

Peripheral vascular disease and other causes of impaired local circulation predispose to cold injury.

Non-freezing cold injury

- **Trench foot   Chilblain**

Trench foot (immersion foot)

Trench foot is caused by prolonged immersion of the feet in cold water or in wet boots, generally at temperatures of 0–10 °C. Intense vasoconstriction results in tissue ischaemia and neurovascular damage without ice crystal formation.

The feet are initially cold and numb and walking is difficult due to the loss of sensation. The skin is pale and mottled and pedal pulses may be impalpable.

If the feet are cold and pulseless they should be rewarmed in a warm bath (37–41 °C). Elevation reduces the oedema and weight bearing must be avoided. After rewarming, the feet become hyperaemic, swollen, and intensely painful. Pain relief is usually required. Ulceration, secondary infection, and even gangrene can occur during this hyperaemic phase, which may last up to six weeks.

Table 2.1 • Wind-chill index

Wind speed (mph)	Equivalent chill temperature (°C)								
0	4	−7	−12	−18	−23	−29	−34	−40	−46
5	2	−9	−15	−21	−26	−32	−37	−43	−48
10	−1	−15	−23	−29	−37	−34	−51	−57	−62
15	−4	−21	−29	−34	−43	−51	−57	−65	−73
20	−7	−23	−32	−37	−46	−54	−62	−71	−79
25	−9	−26	−34	−43	−51	−59	−68	−76	−84
30	−12	−29	−34	−46	−54	−62	−71	−79	−87
35	−12	−29	−37	−46	−54	−62	−73	−82	−90
40	−12	−29	−37	−48	−57	−65	−73	−82	−90

Winds above 40 mph have little additional effect.

Little danger. | Increasing danger. Flesh may freeze within one minute. | Great danger. Flesh may freeze within 30 s.

10 mph = 16.1 km/h
Data redrawn from Mills, W. J. (1973). Frostbite and hypothermia. Current concepts. *Alaska Medicine*, **15**, 26–59.

Eventually the pain and swelling resolve and sensation usually returns. However, there may be muscle wasting and hypersensitivity of the feet to cold, and pain on walking may persist for years.

Chilblain (perniosis)

Chilblains are skin lesions primarily occurring on the dorsal surface of hands and feet. The injury is characterized by localized erythema and swelling. In severe cases, cutaneous haemorrhagic vesicles and ulceration can occur. The skin lesions appear 12–14 hours following cold exposure, particularly in damp climates. Chilblains are most frequently seen in young women and in individuals with a predisposing condition such as Raynaud's disease. Patients complain of intense pruritus and burning paraesthesiae.

The underlying pathology is a dermal lymphocytic vasculitis associated with microvascular stasis, thrombosis, and oedema precipitated by a vascular sympathetic hypersensitivity to cold.

The affected areas should be elevated to reduce the oedema, cleaned, and dressed. Tender blue nodules appear on rewarming and can persist for weeks. Healing may be followed by hyperpigmentation of affected areas and victims are prone to recurrences after milder exposure.

Box 2.2 Cold injuries

Non-freezing — Trench foot
— Chilblain

Freezing Frostnip
Frostbite
— Superficial
— Deep

Pathophysiology of frostbite

- **Vasoconstriction   Ice crystal formation   Endothelial injury   Role of prostaglandins and thromboxane A_2**

The initial response to tissue cooling is arteriolar vasoconstriction. Decreased capillary perfusion causes sludging and thrombus formation. Multiple arteriole-to-venous shunts then develop which open and close in cycles in an attempt to preserve the viability of the extremity. Ultimately, when the core temperature is at risk the shunt cycles cease and the tissue freezes.

Generally, freezing occurs over a period of hours which results in ice crystal formation from *extracellular* free water. The remaining hypertonic extracellular fluid causes osmotic intracellular dehydration. Loss of cellular fluid causes pH and electrolyte imbalance and disruption of enzymatic activity. These changes may ultimately result in cell death although a window of survival is available given appropriate treatment. When rapid freezing occurs within seconds or minutes *intracellular* ice crystals form and immediate cell death results.

During tissue freezing the vascular endothelium may be disrupted. The loss of microvascular integrity can subsequently lead to tissue damage and necrosis during thawing and rewarming.

When the affected tissue is rewarmed the ice crystals melt. Immediately following thawing there is complete restoration of circulation. However, within minutes, erythrocytes and platelets aggregate and thrombi develop which leads to microvascular stasis. The stasis is precipitated by leakage of fluid and protein as the damaged endothelial lined capillaries dilate. The tissue becomes more oedematous as reperfusion continues. Progressive dermal ischaemia during reperfusion may be compounded by release of prostaglandins and thromboxane A2 from damaged endothelial cells which cause leucocyte and platelet aggregation and vasoconstriction.

Final demarcation between viable and non-viable tissue can take 2–3 months. Following severe frostbite injury, arterioles and venules disappear from the microvasculature of the affected tissue. If necrosis does not supervene a distorted microangiographic pattern remains.

Clinical features of frostbite

- **Frostnip   Superficial vs. deep frostbite**

Frostbite most commonly involves the fingers, toes, nose, and ears.

Frostnip may precede frostbite. The affected skin is white and numb. On rewarming paraesthesiae and hyperaemia occur briefly but there is no tissue loss.

Superficial frostbite involves the skin and subcutaneous tissue. The frozen area has a white waxy appearance and is numb. Although the tissue may feel hard the deep tissue is still pliable. After thawing the frostbitten area becomes hyperaemic and very oedematous. During rewarming the skin becomes mottled or purple and large serum-filled blisters develop within 24 hours. The blisters resorb or rupture leaving a hard black eschar. There is usually a throbbing burning pain which lasts for several weeks. When the eschar separates after 3 to 4 weeks delicate sensitive red shiny skin is revealed beneath. The area often sweats excessively.

Deep frostbite involves muscle, nerves, and sometimes bone as well as skin and subcutaneous tissue. The injured area cannot be depressed and feels wooden. Despite rewarming the area remains blue-grey or marble-white in appearance with little associated swelling. Small dark haemorrhagic blisters may appear at a later stage along the line of demarcation between viable and dead tissue. The viable tissue eventually retracts from the mummified part which separates after several months.

At the scene

Frostnip is the only form of cold injury which should be treated in the field. Signs of frostnip on the face or nose on other members of an expedition party should be carefully watched for. If whitening of the skin develops, immediate shelter from the wind should be sought and the affected area warmed by the hand. When normal skin colour returns climbing and walking can resume.

Refreezing of a frostbitten extremity which has been rewarmed in the field results in extensive tissue loss. **There should be no attempt to actively rewarm the affected part at the scene or during transport.** Slow partial rewarming near camp fires or stoves should be avoided as thermal burns may occur to the frostbitten tissue which is devoid of sensation.

The extremity should be padded and splinted for protection. All wet and constricting clothing must be removed and the patient insulated against further heat loss. Rubbing the affected area with a warm hand or snow does not improve blood supply and may risk damaging the skin and introduce infection.

If caught in cold weather with frostbitten feet it is advisable to continue walking in an attempt to reach safety rather than rewarm the feet and then return to the cold. No patient should be allowed to walk on thawed or partially thawed feet.

Immediate treatment in the Accident and Emergency department

Systemic hypothermia should be treated before management of the frostbite injury is started.

Rapid rewarming of the frozen area in a water bath at 37–41°C should be commenced immediately. The temperature should be checked frequently and hot water added carefully and thoroughly mixed. Rewarming can usually be stopped after 20–30 minutes when hyperaemia heralds the return of circulation and the tissue is soft and pliable. Injured extremities should not be rewarmed with dry heat as it is difficult to regulate and can cause burns. Rewarming is not indicated if the area is thawed by the time the patient presents to the Accident and Emergency department.

Thawing is accompanied by intense pain and titrated opiate analgesia is often required.

After rewarming the extremity should be enclosed in a 'burns bag' and elevated. Protection by a cradle prevents disruption of blisters.

Daily cleaning in a whirlpool bath containing antiseptic is useful during which active exercise of the extremity should be encouraged. A whirlpool effect can be created by bubbling oxygen through the water.

Superficial white or clear blisters contain prostaglandins and should be aspirated, repeatedly if necessary, but the skin should be left to protect underlying tissue. Deep haemorrhagic blisters should be left intact as debridement can cause desiccation of the underlying dermis and conversion of the superficial injury to a deep injury.

An antitetanus toxoid booster is required if prophylaxis is not up to date.

Oral ibuprofen, which inhibits prostaglandin synthesis, has been shown to improve outcome. Smoking should be prohibited because of the vasoconstriction induced by nicotine.

The black eschar acts as a protective covering for the regenerating tissue and will usually separate by itself. Attempts to aid separation should be avoided as infection and greater tissue loss may result. Escharotomy may be required to permit movement and prevent flexion contractures. **However, surgical debridement and excision is not indicated for several months until spontaneous amputation is complete.**

All patients should be warned of the increased susceptibility to frostbite on future exposure to cold.

Further reading

Bangs, C. C. (1984). Hypothermia and frostbite. *Emergency Medicine Clinics of North America*, **2**, 475–87.

Britt L. A., Dascombe W. H., and Rodriguez, A. (1991). Management of hypothermia and frostbite injury. *Surgical Clinics of North America*, **71**, 345–70.

Heggers, J. P., Robson, M. C., and Manavalen, K. (1987). Experimental and clinical observations on frostbite. *Annals of Emergency Medicine*, **16**, 1056–62.

McCauley, R. L. *et al.* (1983). Frostbite injuries: a rational approach based on the pathophysiology. *Journal of Trauma*, **23**, 143–7.

Ward, M. (1974). Frostbite. *British Medical Journal*, **1**, 67–70.

Washburn, B. (1962). Frostbite: What it is—how to prevent it—emergency treatment. *New England Journal of Medicine*, **266**, 974–89.

CHAPTER 3

Near drowning

General introduction	41
Pathophysiology	42
At the scene	44
Aspects of resuscitation	45
Complications of near drowning	49
Giving a prognosis	50
Further reading	50

Key points in near drowning

1. Death should never be declared at the scene unless evacuation to hospital is impossible or exhaustion of the rescuer supervenes.
2. The initial management of near-drowning victims is identical irrespective of the type of water involved. Most victims reaching hospital have not aspirated sufficient salt water or fresh water to cause haemodynamic or electrolyte disturbances.
3. Despite a history of prolonged submersion, resuscitation should be continued until hypothermia and drugs have been excluded as far as is practicable as a cause for coma.
4. All patients must be admitted to hospital for observation because of the danger of secondary drowning from the adult respiratory distress syndrome.
5. One of the most important prognostic indicators is the time to the first spontaneous gasp.

General introduction

- **Terminology   Hypothermia   Precipitating factors**

Immersion accidents result in 700 deaths from drowning every year in the United Kingdom. The frequency of non-fatal accidents is, however, much greater. Such accidents are liable to remain one of the leading causes of accidental deaths whilst participation in aquatic activities continues to increase. The majority of drownings occur in inland waters. Coastal waters are associated with better surveillance and rescue procedures.

There are several commonly used terms which refer to immersion accidents. *Drowning* is defined as suffocation from submersion in a liquid and *near drowning* implies that recovery has occurred, at least temporarily, from this. Following recovery, symptoms and signs of the adult respiratory distress syndrome may develop within a few hours after aspirating water and is sometimes referred to as *secondary drowning*.

In 85 per cent of drowning and near-drowning victims, fluid is aspirated into the lungs. *Dry drowning* refers to the remaining 15 per cent of cases caused by asphyxiating laryngospasm precipitated by the initial entry of water into the larynx.

Immersion incidents may be complicated by hypothermia. With the increasing use of life-jackets protecting the victim from submersion, hypothermia has become an important cause of death. In very rough seas, with water continually breaking over the face of a supported victim, drowning can occur without complete submersion when consciousness becomes impaired from supervening hypothermia. Exposure to cold water causes reflex hyperventilation which increases the risk of aspirating water. Near-drowning cases presenting to Accident and Emergency departments in the United Kingdom are often associated with hypothermia which requires the appropriate active intervention outlined in Chapter 1.

A variety of factors may precipitate a submersion accident. The single most important factor in adult drownings is alcohol consumption. Injury to the cervical spine can occur after diving or falling into shallow water. An awareness of these and other confounding factors must accompany the assessment of near-drowning victims (Box 3.1).

> **Box 3.1 Predisposing factors in submersion accidents**
> - Alcohol
> - Psychomotor depressive drugs
> - Poor swimming ability
> - Injury
> - Seizures
> - Myocardial infarction
> - Cerebrovascular accident

The optimum management of near-drowning victims includes proficient rescue, skilled resuscitation at the scene, expeditious evacuation to hospital, advanced cardiopulmonary resuscitation in the Accident and Emergency department, intensive care therapy and monitoring, and making realistic predictions about the prognosis in survivors.

Pathophysiology

- **Pulmonary injury** Sea water vs. freshwater drowning

Pulmonary injury

Involuntary submersion initially results in voluntary breath-holding. A breakpoint is eventually reached resulting in involuntary inspiration and aspiration. Breath-holding time is one-third normal at water temperatures of less than 15°C. In very cold water there is reflex hyperventilation and the likelihood of unintentionally aspirating water is greatly increased. Pulmonary injury and accompanying hypoxia may occur following aspiration of only 1–3 ml/kg of water into the lungs. In both salt and fresh water aspiration a number of factors contribute to the pulmonary insult (Box 3.2).

> Box 3.2 **Pulmonary injury in near drowning**
> - Decreased surfactant — salt water — denaturation
>   — fresh water — loss from washout
> - Increased airway resistance
> - Alveolar capillary membrane damage
> - Reflex pulmonary vasoconstriction
> - Interstitial and alveolar pulmonary oedema
> - Atelectasis

These pulmonary insults result in intrapulmonary shunting which leads to profound hypoxia. Contaminants such as mud, sewage, and bacteria will influence the pattern of pulmonary insult. It is not known if chlorinated fresh water increases the severity of pulmonary injury.

The victim then enters a decompensation phase with further gasping and inhalation. Substantial volumes of water may be swallowed during this phase. The hypoxia becomes more profound and a metabolic acidosis follows, compounding an already established respiratory acidosis. Increasing hypoxia leads to a breakdown of the blood–brain barrier and consequent central nervous system dysfunction. The extent of neurological impairment will depend on confounding factors such as the patient's age, drug and alcohol ingestion, and presence of hypothermia.

Both hypoxia and acidosis lead to cardiac arrhythmias, the development of which is also modified by hypothermia.

Sea water vs. fresh water aspiration

The pathophysiological differences between fresh water and sea water aspiration have received considerable attention. Sea water has an osmolality 3–4 times that of plasma. Therefore, in theory, aspiration of hypertonic sea water should cause fluid shifts from the intravascular space to alveoli, a decrease in blood volume, and an increase in most electrolytes. The reverse of such changes should be seen in cases of

44 • Near drowning

fresh water drowning. These differences have been observed in experimental studies but most victims of near drowning presenting to hospital have not aspirated enough fluid to cause significant haemodynamic or electrolyte changes.

At the scene

• **Rescue and resuscitation**   **Transfer to hospital**

Rescue and resuscitation

The initial management in sea water and fresh water submersion accidents is identical. At the scene resuscitation should follow basic first aid principles.

Expired air resuscitation can be initiated while the victim is still in the water but external cardiac compressions cannot be delivered adequately until the victim is rescued from the water. In near-drowning accidents associated with sport diving, scuba regulators have been used for assisted ventilation. During expired air ventilation the pressure required to inflate the lungs may be considerable. Lung compliance is dramatically reduced after aspiration of salt or fresh water and especially in cases where the inhaled fluid contains paint, fertilizer, or sewage.

Following submersion accidents from surfing and diving into water, cervical spine immobilization should be maintained during rescue and resuscitation.

There is a high risk of further aspiration secondary to regurgitation of large volumes of water and other stomach contents. The unconscious self-ventilating patient should be placed and transported in the coma position. A modified position avoiding flexion or rotation of the neck must be used if cervical spine injury is suspected.

Attempts to drain fluid by gravity from the respiratory tract or stomach should be avoided because they may delay cardiopulmonary resuscitation and exacerbate a cervical spine injury.

Adequate ventilation will often restore satisfactory cardiac function. However, external cardiac compressions should be commenced if a major pulse cannot be palpated after the first few inflations. The pulse can be difficult to feel in hypothermic victims and careful palpation is needed as intervention when the heart is still beating may precipitate ventricular fibrillation.

Following rescue, a further drop in core temperature should be minimized by adequate insulation of the patient. The patient should be carried flat.

When suitable equipment is available, high flow oxygen via a mask should be administered to spontaneously breathing patients. Apnoeic patients require assisted ventilation which is greatly facilitated by skilled tracheal intubation. Defibrillation should be attempted in the field but in severe hypothermia it may not be successful until core rewarming has raised the body temperature. Often, the first sign of successful cardiopulmonary resuscitation is a convulsive abdominal diaphragmatic heave with a flood of vomit or swallowed water.

Transfer to hospital

Prolonged submersion of up to 40 minutes may be compatible with complete recovery even when cardiopulmonary arrest is initially present. Accurate assessment of the situation at the scene is extremely difficult and it is recommended that resuscitation is continued until the patient is transferred to hospital. Death must never be declared at the scene if the victim is cold, unless rescuer exhaustion supervenes and evacuation to hospital is impossible.

Aspects of resuscitation

- History   Investigations   Assisted ventilation
  Advanced cardiac life-support   Hypothermia   Acid–base disturbances   Other considerations

The clinical history is almost always incomplete during the initial phase of management. Important information can be obtained from bystanders, rescuers, police, and ambulance personnel. An ideal history will contain those factors listed in Box 3.3.

> **Box 3.3 Important factors in the clinical history**
>
> - Time of accident
> - Time of rescue
> - Duration of immersion/submersion
> - Water temperature
> - Quality of water — Clean or contaminated
>   — Salt or fresh
> - Skill of initial resuscitation
> - Aspiration of vomit
> - Time to first spontaneous gasp
> - Past medical history
> - Special features of incident

Investigations

Initial investigations which should be performed in all victims of submersion accidents are given in Box 3.4.

> **Box 3.4 Initial investigations in submersion victims**
>
> - Arterial blood gases — should not be corrected for core temperature
> - Urea and electrolytes
> - Blood glucose
> - Chest X-ray
> - Core temperature
> - 12 lead electrocardiogram

In the Accident and Emergency department resuscitation will depend on the patient's overall clinical condition but should be guided by arterial blood gas analysis and the patient's conscious level. Treatment of near-drowning victims is aimed at the restoration of adequate ventilation and circulation, insulation against further heat loss and

Near drowning • 47

rewarming as appropriate, and the correction of acid—base abnormalities.

Assisted ventilation

High flow oxygen should be administered via a face mask to all conscious patients while awaiting the results of arterial blood gas analysis.

Clearing the airway of regurgitated fluid and debris by suctioning is of primary importance in unconscious casualties. If the gag reflex is significantly impaired or the patient is not breathing, the airway should be secured by skilled tracheal intubation. If such skills are not available, ventilation should be maintained by bag and mask and appropriate help summoned.

Mechanical ventilation is required when a patient cannot maintain satisfactory arterial oxygen and carbon dioxide concentrations. Intermittent positive pressure ventilation should be commenced in the resuscitation room if the following occur:

- PaO_2 of 8.0 kPa (60 mm Hg) or less breathing air
- PaO_2 of 10.0 kPa (75 mm Hg) or less on 50 per cent inspired oxygen
- $PaCO_2$ of 7.5 kPa (50 mm Hg) or more
- Signs of pulmonary oedema

Advanced cardiac life-support

If cardiac arrest occurs, advanced life-support measures should be initiated and cardiac arrhythmias treated according to standard protocols. In patients with reduced core temperatures the guidelines for the management of hypothermic victims presented in Chapter 1 should be followed. Although it has been suggested that hypothermia may offer cerebral protection there is little data to support the use of induced hypothermia in near-drowning cases and it is not recommended.

Hypothermia

Cardiopulmonary resuscitation should not be abandoned until the core temperature exceeds 33 °C and drugs have been excluded as a cause of the arrest. When metabolic and thermal imbalances have been corrected and resuscitation has continued for an additional 30 minutes in the presence of asystole without response, it is usually reasonable to stop.

Acid–base disturbances

Acidosis is the most common and most serious metabolic abnormality encountered in submersion victims. Initial management is directed towards restoration of adequate ventilation. In prolonged resuscitation, 50 mmol of bicarbonate may be given to an adult patient, but further aliquots should be guided by arterial blood gas values.

Other considerations

Serum electrolyte disturbances are not common and if present usually resolve following adequate resuscitation and rewarming.

Substantial volumes of water may be absorbed from the stomach especially in infant victims and pulmonary aspiration causes severe complications. A nasogastric tube should therefore be passed. Patients with facial injury or in whom there is suspicion of a cribriform plate or basal skull fracture should have an orogastric rather than a nasogastric tube inserted to avoid inadvertent intracranial placement.

Physical examination may reveal associated injuries. There should be a high index of suspicion for cervical spine injury. Whenever there is uncertainty of the circumstances surrounding the incident and especially if the conscious level is depressed, the casualty should be assumed to have an unstable cervical spine injury. A lateral cervical spine X-ray must be performed in these circumstances.

A chest radiograph must be performed in all near-drowning cases. The typical radiographic pattern is generalized pulmonary oedema with patchy shadowing. The radiograph may be normal despite a pulmonary insult and the initial appearances often under-estimate the extent of the pulmonary injury.

Prophylactic antibiotics need not be administered as a routine to all submersion victims. Antibiotic treatment is usually restricted to patients with confirmed sepsis on sputum or blood culture but antibiotic therapy should probably be started early if contaminated water such as sewage was involved. Steroids have not clearly been shown to improve the outcome following submersion injury and their administration is not recommended.

Positive end expiratory pressure (PEEP) ventilation typically results in marked improvement in the patient's oxygenation. PEEP increases the functional residual capacity, improves ventilation–perfusion mismatch by reducing atelectasis, and enhances fluid resorption from the pulmonary interstitium. Up to 25 cm H_2O of PEEP may be required to achieve satisfactory oxygenation of blood. Such pressures decrease venous return to the heart and vigorous expansion of plasma volume is often needed. Prompt admission to an appropriate intensive therapy unit must be arranged.

Complications of near drowning

Up to 15 per cent of near drowning victims who are conscious at the time of admission to the Accident and Emergency department subsequently die as a result of pulmonary or cerebral complications. All patients presenting to hospital following submersion accidents require admission for a period of observation.

Box 3.5 Complications of near drowning

- Adult respiratory distress syndrome
- Cerebral oedema
- Renal failure
- Disseminated intravascular coagulation
- Sepsis

Clinical features of secondary drowning from the adult respiratory distress syndrome usually develop within 4 hours after the submersion accident. Asymptomatic patients with normal serial arterial blood gas results while breathing air and chest X-ray after 6 hours are unlikely to develop problems and can be safely discharged.

Giving a prognosis

The management of submersion victims includes providing a realistic expectation of outcome to relatives. The duration of submersion and hence hypoxia, which is the most important variable, is usually unknown.

Immersion in cold water and resulting hypothermia affords protection of the brain. Circulatory arrest resulting from ventricular fibrillation in children occurs at a lower temperature than in adults thus permitting profound cerebral cooling. This may explain the successful resuscitation from prolonged cold water submersion episodes in children and infants.

Prompt and skilled basic cardiopulmonary resuscitation at the scene is associated with a better chance of recovery.

Patients transferred from the Accident and Emergency department to an intensive therapy unit and who arrive comatose with fixed dilated pupils have a poor prognosis.

Probably the best prognostic indicator is the time to the first spontaneous gasp, which if it occurs within 30 minutes of rescue, is associated with a good outcome.

Further reading

Golden, F. St. C. and Rivers, J. F. (1975). The immersion incident. *Anaesthesia*, **30**, 364–73.

Harries, M. (1986). Drowning and near drowning. *British Medical Journal*, **293**, 122–4.

Kram, J. A. and Kizer, K. W. (1984). Submersion Injury. *Emergency Medicine Clinics of North America*, **2**, 545–52.

Pearn, J. (1985). The management of near drowning. *British Medical Journal*, **291**, 1447–52.

Stuart Taylor, M. E. (1990). Management of near drowning. *Hospital Update*, **16**, 419–31.

CHAPTER 4

Diving emergencies

General introduction	53
Diving physics	54
Pressure	54
Gas laws	54
Barotrauma of descent	57
Problems encountered at depth	58
Pulmonary barotrauma of ascent	59
Decompression sickness	60
At the scene	63
Management in the Accident and Emergency department	65
Further reading	67

Key points in diving emergencies

1 Symptoms developing within 48 hours of a dive should be regarded as due to a dive-related illness until proved otherwise. The problem must be discussed immediately with a diving specialist.
2 Urgent therapeutic recompression is mandatory for cerebral air embolism and decompression sickness.
3 All divers with serious dysbaric illness should be given 100 per cent oxygen. This promotes the maximum gradient of nitrogen gas excretion and improves oxygenation of ischaemic tissues.
4 Analgesia is contraindicated as it will mask the response of the patient's symptoms to recompression.
5 Administration of Entonox to a diving casualty is contraindicated as nitrous oxide will diffuse into and enlarge nitrogen bubbles.
6 Diving equipment and the casualty's diving 'buddy' should accompany the patient to the recompression chamber.

General introduction

- **Epidemiology Dysbarism Self-contained underwater breathing apparatus**

It is estimated that there are about 50 000 sport divers in the United Kingdom performing 1–2 million dives every year. In addition, 1500 commercial divers operate in the North Sea. The medical care of these professional divers is strictly controlled by law and supervised by a few diving medicine specialists. Commercial divers are supported by a skilled surface team and accidents are now rare. Although there is no legislation controlling recreational diving, national sub-aqua clubs promote high standards of training and safety. Despite this 10–12 deaths and approximately 200 serious diving accidents are reported every year. Not only has there been an upward trend in the number of serious casualties but also the proportion of sport divers presenting to hospitals with neurological manifestations of diving illness has increased.

The medical problems of sport diving are primarily due to intrinsic hazards of the aquatic environment and to breathing compressed air at greater than atmospheric pressure. The pressure-related problems that are associated with diving are collectively known as *dysbarism*.

Doctors and paramedics may have to start resuscitation of a diving casualty at the scene and arrange evacuation to a specialist (diving) medical facility. Furthermore, divers may present to Accident and Emergency departments far removed from the dive site with vague symptoms which may herald serious dysbaric illness. The consequences of delay in diagnosis and appropriate management can be catastrophic.

Self-contained underwater breathing apparatus (SCUBA) used by sport divers consists of a compressed air cylinder carried on the back. The pressure of air in the cylinder needs to be reduced before the diver can breath it. This is achieved by a regulator and in the most popular type, the pressure is reduced in two stages. The pressure is first decreased by a reduction valve which is attached to the top of the cylinder.

The air at intermediate pressure is then fed via a hose to a demand valve attached to a soft rubber mouthpiece. The demand valve ensures that the pressure of air breathed by the diver is exactly equal to the surrounding water pressure. Sport divers use SCUBA at pressures associated with depths down to 50 metres.

Diving physics

- **Pressure   Gas laws**

Divers encounter very different environmental conditions including changes in visibility, sound conduction, temperature, density, and ambient pressure. The most important contributing factor in serious diving-related injury and illness is change in pressure. An appreciation of basic pressure physics and physiology provides a clearer picture of the medical problems which can arise from diving and the principles involved in their management.

Pressure

The pressure exerted by air contained in the atmosphere at sea level is approximately $1\,kg/cm^2$ (1 atmosphere) although we are generally unaware of this. When a diver descends underwater the ambient pressure increases. Water is much denser than air and the pressure exerted by 10 metres of sea water is the same as that exerted by the entire atmosphere above sea level. The absolute pressure at this depth is therefore 2 atmospheres. Sport diving is performed at pressures of up to 6 atmospheres.

The tissues of the body are composed mostly of water, which is not compressible and not significantly affected by pressure changes. However, gases are compressible and consequently gas-containing spaces and organs are directly affected by changes in ambient pressure.

Gas laws

The basic mechanisms of dysbaric diving disorders are explained by three laws of physics.

Diving emergencies

Table 4.1 • Boyle's law

	Depth (ms)	Absolute pressure (atms)	Gas volume (%)
Air	0	1	100
Sea water	10	2	50
	20	3	33
	30	4	25
	40	5	20
	50	6	17

Boyle's law. The volume of a given mass of gas is inversely proportional to the absolute pressure if the absolute temperature remains constant (Table 4.1). It can be seen that the greatest changes in pressure ratio and consequently volume occur near the surface.

If a person fills their lungs at the surface and dives to 10 metres, the lung volume will decrease by 50 per cent. On returning to the surface the gas will re-expand to its original volume. When SCUBA is used during a dive, the pressure in the gas-filled spaces of the body is normally in equilibrium with the ambient pressure. If the volume of gas spaces in the diver's body is to remain constant during a descent, air has to be added as the ambient pressure increases (Fig. 4.1). Should obstruction occur in the various portals of gas exchange a pressure disequilibrium develops. During the ascent phase the ambient pressure decreases and gas within the body spaces expands in volume. Unless the gas is vented as it expands it will exert pressure on the surrounding tissues (Fig. 4.1). Injury caused by direct tissue damage following volume change of gas consequent on pressure is called *barotrauma* and can occur during *descent* or *ascent*. Barotrauma can occur during (usually rapid) ascent following a shallow dive—even from the bottom of a swimming pool.

Dalton's law. Air consists of approximately 21 per cent oxygen, 79 per cent nitrogen, 0.03 per cent carbon dioxide, and traces of other gases. The total pressure of this mixture of gases is equal to the sum of their partial pressures (Dalton's law). The biological effects of each gas depend on its partial pressure which changes in proportion to the

56 • Diving emergencies

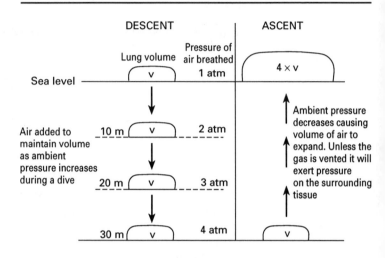

Fig. 4.1 • Pressure/volume relationships during a SCUBA dive.

ambient pressure. The significance of partial pressure in diving concerns the toxic effects which various gases have on the body at elevated pressure: 100 per cent oxygen cannot be used in diving because of the associated toxicity when breathed at a partial pressure in excess of 2 atmospheres absolute. The partial pressure of nitrogen in air increases with increasing depth, and below about 30 metres exerts a narcotic effect (*nitrogen narcosis*), which increases with increasing depth.

Henry's law. The amount of gas dissolved in a given volume of fluid is proportional to the pressure of gases upon that fluid (Henry's law). At sea level a diver's body tissue contains about 1 litre of gaseous nitrogen in solution. At a depth of 10 metres breathing air at 2 atmospheres absolute a diver will eventually reach equilibrium again and have twice as much nitrogen dissolved in the tissues. When a diver then ascends to the surface the extra nitrogen load must be expelled in a controlled fashion. If this does not occur desaturation of dissolved nitrogen forms bubbles in tissues and blood, resulting in *decompression sickness*.

Barotrauma of descent

- **Aural, sinus, and skin barotrauma**

All gas-filled compartments are subject to barotrauma of descent (and ascent). Barotrauma of descent or 'squeeze' as it is referred to by divers, results from compression of gas in enclosed spaces as ambient pressure increases. Aural barotrauma is the most common type of injury.

Aural barotrauma

Equalization of middle ear pressure can be facilitated during a diver's descent by swallowing or performing a Valsalva manoeuvre. Mucosal congestion may obstruct the eustachian tube and as the pressure differential increases the diver will experience discomfort. As expected from the way in which gas volume changes with depth, most middle ear squeezes occur near the surface. If the diver continues to descend the pressure differential increases, mucosal oedema and haemorrhage occur, and *inward* bulging of the tympanic membrane may eventually lead to rupture. Cold water then enters the middle ear causing vertigo and disorientation which can precipitate a diving accident. Clinical examination will show erythema, haemorrhage, or perforation of the tympanic membrane associated with a conductive hearing loss.

External ear squeeze results if the external auditory canal is occluded by, for example, cerumen plugs. Compression of the enclosed air will result in *outward* bulging of the tympanic membrane, and an effect on the lining of the external auditory canal similar to that of the middle ear cavity in middle ear barotrauma of descent.

Rupture of the round or oval window ('inner ear squeeze') can occur if sudden and marked pressure changes occur between the middle and inner ear. A perilymphatic fistula results and classically the patient complains of tinnitus, vertigo, and deafness. The typical findings of middle ear barotrauma, sensorineural hearing deficit, and vestibular dysfunction are found on examination.

If tympanic membrane rupture has not occurred, middle ear squeeze can be managed with a decongestant, which usually results in resolution of symptoms within two weeks. Antibiotics should be prescribed when there is tympanic membrane rupture, pre-existing infection or following a dive in polluted waters. The diver must refrain from diving until the symptoms resolve and a rupture has completely healed.

Specialized otolaryngological opinion is urgently required if perilymphatic fistula is suspected and also for follow-up in cases of severe or continuing symptoms.

Sinus barotrauma

Paranasal sinuses may be similarly affected by the development of pressure differentials. Precipitating factors include upper respiratory tract infections, mucosal polyps, and sinusitis. Treatment is similar to aural barotrauma.

Skin barotrauma

If the diver fails to exhale via his nose into his mask periodically during descent, facemask squeeze can result with erythema, ecchymosis, and petechial haemorrhages of the enclosed skin. Areas of skin tightly enclosed by part of a diving suit can result in similar appearances. These injuries appear spectacular but usually no treatment is required.

Problems encountered at depth

- **Nitrogen narcosis**

The narcotic effects exerted by nitrogen at elevated partial pressures are similar to those of alcohol intoxication. Nitrogen is very fat-soluble and dissolves in the lipid component of nerve cell membranes slowing impulse conduction. At depths below 30 metres euphoria, over-confidence, and confusion become significant and decreasing conscious level below 50 metres renders air diving unsafe. In commercial diving, helium is used below this depth to dilute the oxygen as helium has no narcotic effects.

There is considerable variation in the symptoms experienced by different divers at the same depth or by a single diver on different occasions. Susceptibility to nitrogen narcosis is increased by recent alcohol or sedative drug ingestion, and by exertion and cold temperatures experienced during the dive. Although the narcotic effects are rapidly and entirely reversed as the diver ascends, nitrogen narcosis can be a precipitating factor in diving accidents. The diver's memory of events surrounding an accident may also be inaccurate and this should be taken into account when taking the history.

Pulmonary barotrauma of ascent

- **Mediastinal emphysema   Pneumothorax   Air embolism**

Barotrauma of ascent is the reverse process of squeeze. It is unusual for the ear or sinuses to be affected by barotrauma of ascent because obstruction of the portals of gas exchange is unlikely if pressure equilibration can be achieved during descent. 'Reverse squeeze' of the middle ear may result if the diver treats upper respiratory congestion with a short-acting topical vasoconstrictor and the effect wears off during the dive.

If the diver does not exhale adequately during ascent, alveoli may rupture allowing air to escape through the visceral and mediastinal pleura or alternatively into the pulmonary capillaries. *Pulmonary barotrauma* may be precipitated by breath-holding during a rapid and uncontrolled ascent or by air trapping which occurs in patients with asthma or congenital bullae.

The main consequences of pulmonary barotrauma are mediastinal emphysema, pneumothorax, and air embolism.

Mediastinal emphysema

Medistinal emphysema is the most common sequela of 'burst lung' and presents with gradually increasing hoarseness, neck fullness, and retrosternal chest discomfort over several hours after diving. Symptoms usually resolve spontaneously or following treatment with high flow oxygen.

Pneumothorax

Pneumothorax is unusual although it is potentially a very serious complication if it develops during a dive. The intrapleural gas cannot be vented and will progress to a tension pneumothorax as the diver ascends. All patients suspected of pulmonary barotrauma should have a chest X-ray to exclude a pneumothorax. Treatment by needle aspiration or insertion of a chest drain is usually required.

Air embolism

Dysbaric air embolism follows entry of air bubbles into the pulmonary capillaries from ruptured alveoli. The bubbles are carried to the left side of the heart and subsequently disseminated through the systemic circulation. Cerebral air embolism causes symptoms as soon as the diver surfaces. Classically, the onset is dramatic with loss of consciousness, convulsions, cardiovascular collapse, and chest pain from the associated pulmonary barotrauma. The motor and sensory deficits tend to be unilateral and unimodal in contrast to the signs of decompression sickness which are usually bilateral and bimodal.

Decompression sickness

- **Joint pain ('the bends')   Cutaneous manifestations
  CNS, pulmonary, and vestibular decompression sickness**

As the diver ascends and decompression occurs the dissolved load of nitrogen is not always expelled at a sufficient rate. Desaturation and other factors cause bubbles to form in the tissues and blood, resulting in the multisystem disorder of decompression sickness (Fig. 4.2). This may even occur in divers who have followed the recommendations for safe rates of ascent contained in diving tables. Decompression sickness is more likely to occur in cold water and in divers who are obese or who exercise excessively during diving.

Fifty per cent of divers developing decompression sickness will become symptomatic within the first hour follow-

Diving emergencies

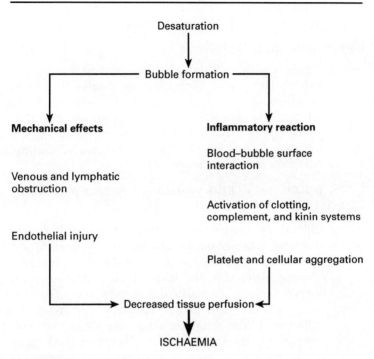

Fig. 4.2 • Pathophysiology of decompression sickness.

ing diving, with 90 per cent developing symptoms within 6 hours. However, patients may still present to the Accident and Emergency department up to 48 hours following a dive, especially after a journey by air.

Joint pain ('the bends')

The colloquial term 'the bends' refers to acute periarticular joint pain and is the most common presenting symptom of decompression sickness. The shoulders and elbows are most commonly affected in air diving decompression sickness. The pain may be described as a dull ache although the diver often finds description difficult. It is usually exacerbated by limb movement but is rarely associated with localized tenderness. Altered sensation may be found around the joint. Inflation of a blood pressure cuff around the joint may temporarily relieve the pain and assist in the diagnosis.

Cutaneous manifestations

Cutaneous manifestations may present as pruritic rashes which may be erythematous or similar to bruises. Local swelling or an orange peel effect can occur if lymphatic obstruction is present. Skin complications other than a transient itch indicate the need for recompression and a careful clinical search for other manifestations of decompression sickness should be made.

CNS, pulmonary, and vestibular decompression sickness

Spinal decompression sickness may result from bubbles (venous, air embolic, or autochthonous) pressing on the cord or obstructing venous drainage in the vertebral venous plexus. Damage to the spinal cord will range from neuropraxia (with early treatment) to infarction. The lower thoracic, lumbar, and sacral segments are most commonly affected. Initial symptoms may be slight, but are always important. Backache, girdle-type abdominal pain, lower limb weakness, and paraesthesiae or urinary retention signify spinal decompression injury.

Other central nervous system involvement gives rise to clinical features which may be focal or diffuse, motor or sensory, or manifest as mood changes.

Cerebellar involvement causes impaired balance, vertigo, and vomiting ('the staggers').

Generalized liberation of nitrogen bubbles in the circulation can result in diffuse pulmonary embolization giving rise to acute respiratory distress ('the chokes'). Fortunately, pulmonary decompression sickness and cardiovascular shock caused by the same mechanisms are rare manifestations.

Bubble formation within the rigid confines of the inner ear can cause disruption of the vestibulocochlear apparatus. Although vestibular decompression sickness is not common, it represents another cause of acute post-dive vertigo. It is important to differentiate between vestibular decompression sickness and inner ear barotrauma as recompression therapy is mandatory in the former and undesirable in the latter.

At the scene

- **Immediate care   Transfer to recompression facility**

Serious diving casualties may require resuscitation prior to transfer to a specialist diving medical centre. However, recompression is the definitive treatment for decompression sickness and cerebral air embolism and any delay will reduce the efficacy of treatment.

Attempts by a symptomatic diver to re-enter the water to perform recompression by descending to the depth of his original dive should be resisted. In UK waters, the risk of developing hypothermia during the time required for such a procedure is considerable and it is generally very hazardous.

Immediate care

All divers developing manifestations of serious dysbaric illness should be given 100 per cent oxygen. This promotes a maximum gradient for nitrogen gas excretion and improves oxygenation of ischaemic tissues.

The left Trendelenberg position previously recommended on theoretical grounds for use in patients suspected of cerebral air embolism has now been identified conclusively as being of no benefit. Additionally, this position makes management procedures technically more difficult and the head-down position may result in an increase in intracranial pressure and exacerbate cerebral oedema.

The pain arising in a joint can often be improved by applying pressure around the joint. The joint may be bound tightly but eased periodically to prevent any distal neurovascular deficit. This symptomatic treatment is not an alternative to recompression.

Analgesics will mask the response to pressure and creates

difficulties for the recompression chamber staff to monitor the response to treatment. Their use is therefore contraindicated.

Development of pain following a recent dive is an absolute contraindication to Entonox administration. If Entonox is inhaled by a diver with decompression sickness, a reverse gradient is created and nitrous oxide will diffuse into and enlarge the nitrogen bubbles. Entonox is also contraindicated in divers injured following a road traffic accident shortly after a dive.

Sedative drugs will also cause difficulty in monitoring the effects of recompression but convulsions associated with cerebral air embolism can be treated by intravenous diazepam titrated against the response. It is important to inform the recompression chamber staff about any such convulsions and anticonvulsant therapy.

Transfer to recompression facility

The history which can be ascertained at the scene and relayed to the diving medical specialist should, if possible, include details of the depth and duration of the last two dives, problems experienced during ascent and descent phases, presence of undue exertion or cold during the dive, omission of any decompression procedures, and any details of non-dive-related illness and medications.

Divers usually dive in pairs, so if one diver has symptoms of decompression sickness or pulmonary barotrauma his 'buddy' will also be at risk of developing dysbaric illness. The buddy should be kept warm and still and, although recompression may not be required, should be transferred along with the affected diver. The diving equipment should accompany both divers to the recompression facility for the staff to inspect.

HM Coastguard (tel. 999) should be contacted from the scene of a serious diving accident to arrange the most appropriate transport to the nearest diving medical centre.

Management in the Accident and Emergency department

With the increasing popularity of the sport and diversity of popular dive sites including inland waters, divers may be brought to Accident and Emergency departments for their initial assessment and treatment. The improvement in diving equipment has allowed recreational divers to undertake deeper and longer dives and may explain the increase in the proportion of divers presenting with neurological features. Additionally, through certified training courses, divers are now more aware of the potential of relatively minor neurological complaints to progress to more serious illness.

The only definitive treatment for severe dysbaric illness is recompression, breathing 100 per cent oxygen. If recompression is delayed, the risk of permanent damage to the brain and spinal cord is greatly increased. *The safest approach is to assume that any symptoms in a diver presenting to the Accident and Emergency department within 48 hours after a dive, are dysbaric in origin.* Advice can be obtained from the Duty Diving Medical Officer on-call at recompression centres (see Appendix).

Often, the diagnosis of decompression sickness may only be confirmed following the response of the symptoms and signs to recompression.

Although the majority of patients presenting to the Accident and Emergency department with dysbaric air embolism will have detectable neurological injury, some patients experience marked resolution of their signs during transfer from the scene. Nevertheless, these patients should be referred to a diving medical centre and recompressed.

The seriously ill diver will require resuscitation prior to transfer to the recompression unit. However, there should be minimal delay in arranging transfer.

High flow oxygen (12 l/min) should be administered to all patients with serious dysbaric illness if this has not already been commenced in the field. If tracheal intubation is required, the cuff should be inflated with sterile water, not air, because during recompression an air-filled cuff will deflate. The recompression staff will always check this.

Intravenous infusion is advocated in serious decompression sickness or air embolism. Normal saline (or a polygeline) may be used. Adequate circulating volume assists oxygenation of the ischaemic tissues and facilitates the discharge of excess tissue nitrogen load into the venous system. Dextran solutions are administered at some diving medical centres in an attempt to prevent capillary sludging which accompanies severe decompression sickness.

Despite using a dry or wet suit for thermal protection, diving related illness is often accompanied by hypothermia. Appropriate passive or active rewarming of the victim should be commenced (Chapter 1).

An indwelling urinary catheter should be inserted in severe decompression sickness or spinal cord air embolism.

High dose parenteral corticosteroids have been advocated to decrease cerebral or spinal cord oedema but should only be administered after discussion with the receiving recompression team. Oral aspirin may be given for its antiplatelet activity.

Transport to the nearest recompression chamber can be arranged in conjunction with the staff of the diving medical centre. If evacuation by air is considered it should be appreciated that the reduction in barometric pressure and partial pressure of oxygen experienced at altitude may exacerbate serious dysbaric illness. Consequently, unpressurized aircraft should not fly above 300 metres and it is particularly important that the diver breathes oxygen. In addition, decompression sickness may be precipitated by air travel in a diver who has left insufficient time between diving and flying for residual nitrogen to leave the body in a controlled fashion. An interval of 12–48 hours is recommended depending on the depth and duration of the dives performed.

On reaching the diving centre, the patient is initially recompressed to a simulated depth of 18 metres of sea water. During treatment, 100 per cent oxygen is administered interspersed with periods of air breathing to reduce the risk of oxygen toxicity. The patient is slowly decompressed according to a standard treatment protocol (modified as necessary).

Further reading

British Sub-Aqua Club (1987). *Sport diving.* Stanley Paul, London.

Douglas, J. D. M. (1985). Medical problems of sport diving. *British Medical Journal,* **291**, 1224–6.

Edmonds, C., Lowry, C., and Pennefather, J. (ed.) (1981). *Diving and subaquatic medicine.* Diving Medical Centre, Australia.

Kizer, K. W. (1984). Diving medicine. *Emergency Medicine Clinics of North America,* **2**, 513–30.

Murrison, A. W. and Francis, T. J. R. (1991). An introduction to decompression illness. *British Journal of Hospital Medicine,* **46**, 107–10.

Sykes, J. J. W. (1990). The hazards of sports diving. *Journal of the Royal College of Physicians of Edinburgh,* **20**, 318–22.

CHAPTER 5

Heat illness

General introduction	71
Heat physiology	72
Acclimatization	75
Minor heat illness syndromes	75
Heat oedema	75
Heat cramps	76
Heat tetany	76
Heat syncope	77
Prickly heat	77
Anhidrotic heat exhaustion	77
Heat exhaustion	77
Heat stroke	79
Clinical features of heat stroke	80
Management of heat stroke	82
At the scene	82
Immediate treatment	83
Methods of cooling	83
Additional measures	84
Prevention	85
Further reading	87

Key points in heat illness

1 In minor heat illness the body's thermoregulatory central mechanisms continue to function. In heat stroke, control is lost and the body temperature rises precipitously causing tissue damage and organ failure.
2 Classical heat stroke typically occurs in epidemic forms during heat waves and primarily affects the elderly and infirm.
3 Exertional heat stroke is seen in young and healthy individuals who generate enormous metabolic heat during vigorous exercise.
4 Adequate fluid intake before, during, and after exercise in hot climates is crucial in avoiding heat-related illness. Supplemental salt tablets are unnecessary.
5 Major central nervous system dysfunction under conditions of heat stress should be considered as heat stroke until proved otherwise.
6 The mortality in heat stroke is primarily related to the duration of hyperthermia and therefore cooling must be started immediately.

General introduction

- **Spectrum of heat illness   Heat stress   Environmental conditions   Precipitating circumstances**

The spectrum of heat illness represents a continuum of disorders which range widely in severity. In minor heat illness syndromes the body's thermoregulatory mechanisms are stressed but continue to function. However, in major heat illness resulting in *heat stroke*, these physiological responses are unable to compensate for the heat stress imposed.

Heat stress represents the combination of heat gain imposed by environmental temperature and exertional heat production, and by the factors which limit or prevent heat dissipation from the body.

Heat illness occurs in situations where individuals at risk, or supervising authorities, do not appreciate the dangers involved during exposure to hot environments. Predisposing factors such as dehydration or lack of acclimatization may precipitate heat illness even in moderate ambient temperatures. Accidental or unforeseeable circumstances may also lead to very high heat stress exposure.

Although most tropical areas have a continuous, albeit varying heat stress throughout the year, there are many other areas in temperate climates where an intermittent seasonal heat stress can be found. Mortality from heat illness increases three-fold in the general population during heat waves, although the elderly and those with cardiovascular disease or other debilitating illness are particularly susceptible. Heat illness is frequently reported amongst unacclimatized Moslems at the Mecca pilgrimage during the hot summer months.

Certain work conditions such as deep mines, engine and boiler rooms, and other locations characterized by high ambient temperatures may contribute to heat illness. The interior of vehicles and tents in the sun provide large radiant heat loads regardless of ambient temperature.

Military personnel are also susceptible to heat illness, especially individuals in the first few weeks of basic training, or during intensive training exercises. Unaccustomed physical activity in hot environmental conditions stresses even young healthy recruits.

With the growing popularity of jogging, running, and cycling more heat illness is being seen. In one UK marathon, 1 per cent of participants were affected despite temperate weather conditions, and cases of fatal heat stroke have occurred during these events.

Heat physiology

- **Metabolic heat production   Heat accumulation and dissipation   Thermoregulation   Fluid loss**

Metabolic heat production

All cellular metabolism produces heat. Basal metabolism would cause a body temperature rise of 1.1 °C per hour in the absence of cooling mechanisms. Active metabolism during physical exercise can increase heat production up to 12-fold. Even with normal cooling mechanisms operating, exercise causes an increase in body temperature. The body's ability to dissipate heat has a fixed maximum rate and with strenuous exercise the endogenous heat load generated may exceed this. Rectal temperatures of 40–42 °C are routine in runners successfully completing marathons.

Heat accumulation and dissipation

The body can also accumulate heat from the environment. Usually this occurs when the ambient temperature exceeds body temperature but it can also occur regardless of air temperature when there is a large radiant heat load. Sitting in bright sunlight may add 150 kilocalories per hour of heat. Similar heat gains occur while working in engine and boiler rooms.

Elevated body temperature imposes an additional intrinsic heat load of its own as cellular metabolism increases approximately 13 per cent for every 1 °C rise in temperature. The core temperature is maintained within a narrow range

Heat illness • 73

by physiological responses which balance heat production and dissipation (Fig. 5.1). There are two main methods by which heat generated from internal and external heat loads is lost. The first is the physical transfer of heat from the body to the cooler environment. Radiation of heat energy from the skin usually accounts for 65 per cent of losses and convection by cooling air currents accounts for 12–15 per cent. Air is a good insulator and only 2 per cent of body heat is lost by conduction. These cooling mechanisms only operate when the ambient temperature is lower than skin temperature (about 32 °C). When air temperature exceeds skin temperature the body accumulates heat.

The second method of heat dissipation is sweating and evaporation, and accounts for virtually all heat loss at ambient temperatures above 32–35 °C. Approximately 1 kilocalorie of heat is lost by vaporizing 1.7 ml of sweat. The average person can produce up to 1.5 litres of sweat per hour, whereas a trained athlete or acclimatized individuals

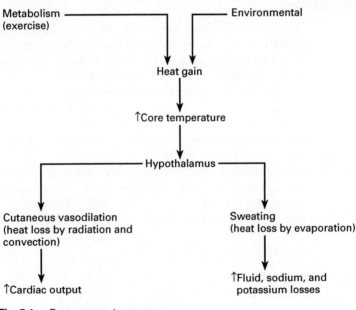

Fig. 5.1 • Response to heat stress.

74 • Heat illness

may produce to 2.5–3 litres of sweat per hour. However, evaporative heat loss is limited by the relative humidity of the surrounding air. As humidity exceeds 75 per cent, evaporation decreases and ceases at levels of 90–95 per cent. The innervation of sweat glands is cholinergic and therefore anticholinergic drugs can be a precipitating factor in developing heat illness.

Thermoregulation

The heat-regulating centre in the hypothalamus monitors core and peripheral temperature and initiates a cardiovascular response in addition to the sweating reflex. Reflex cutaneous vasodilation can increase skin blood 20-fold thereby enhancing heat radiation to the environment. As peripheral resistance falls creating a shunt of approximately 4 litres per minute intense compensatory splanchnic and renal vasoconstriction occur. Cardiac output increases 2–4 times to maintain systemic blood pressure.

Fluid loss

In addition to fluid loss from sweating, intense muscular activity during exercise causes fluid shifts into muscle cells. This occurs in response to the osmotic gradient created by the breakdown of glycogen during exercise. The contraction in intravascular volume develops despite the requirement for increased cardiac output. Ultimately, peripheral vasoconstriction may supervene to maintain systemic blood pressure and consequently heat dissipation is diminished and hyperthermia develops.

Dehydration, particularly when associated with hypernatraemia, can itself increase body temperature by increasing the work of the cellular sodium potassium ATPase pump which accounts for 20–40 per cent of basal metabolic rate.

Acclimatization

- **Renin-angiotensin-aldosterone system   Cardiovascular adaption**

Acclimatization to warm environments is characterized by a lower core temperature threshold for sweating, increased sweat volume, and lowered sweat electrolyte concentration. Cardiovascular and metabolic efficiency improves especially in response to exercise. An increase in the number of mitochondria per cell and muscle glycogen stores enhances aerobic performance. Cardiovascular changes are due to a 10–25 per cent expansion in plasma volume. Early during the acclimatization process, stimulation of the reninangiotensin-aldosterone system promotes renal sodium retention and potassium loss and an increase in total body water. The rise in plasma volume results in a greater stroke volume, slower heart rate, and improved myocardial performance. These adaptive changes occur over several days to weeks during exposure to hot climates. The acclimatized individual, when subjected to equivalent work and heat loads, has a smaller rise in body temperature than unacclimatized people.

Minor heat illness

- **Heat oedema   Heat cramps   Heat tetany   Heat syncope   Prickly heat   Anhidrotic heat exhaustion   Heat exhaustion**

In minor heat illness syndromes, normal temperature regulation is maintained. The symptoms result from the stress of heat dissipation during heat exposure.

Heat oedema

Swelling of the feet, ankles, and hands may develop during the first few days of exposure to a hot environment. It is common in the elderly even in the absence of underlying cardiac, hepatic, or renal disease.

Heat oedema is due to cutaneous vasodilation and mild vascular leak combined with fluid retention. It is self-limiting and usually resolves after several days of acclimatization. Evaluation to exclude deep venous thrombosis or congestive cardiac failure is appropriate but no special treatment is required. Diuretics should not be administered as heat stroke may be precipitated especially in the elderly.

Heat cramps

Heat cramps are intensely painful involuntary spasms of skeletal muscle. They typically occur in heavily exercised muscles after exercise has ceased. Heat cramps occur most commonly in individuals doing strenuous exercise who sweat profusely and replace fluid losses with hypotonic solutions and water. The calf muscles are usually affected although any muscle group can be involved including those of the abdomen. Recent ingestion of alcohol, lack of sleep or inadequate diet, intercurrent disease and lack of acclimatization are all predisposing factors.

Hyponatraemia in the muscle cell is believed to interfere with calcium-dependent relaxation. Although salt and fluid imbalance is postulated as the underlying cause, the serum sodium level is probably not the only determining factor.

Stretching the cramping muscle will quickly relieve symptoms. Treatment of heat cramps consists of salt and fluid replacement and rest in a cool environment. A saline solution of 1 teaspoonful of salt in 500–1000 ml of water is usually tolerated. Salt tablets can cause vomiting and gastrointestinal irritation, as well as an intraluminal hyperosmosis compounding fluid losses and therefore should be avoided.

Severe cases and patients unable to tolerate oral replacement respond to 1 litre of normal saline intravenously over 1–3 hours.

Heat tetany

Heat tetany is a self-limiting condition caused by the respiratory alkalosis which develops secondary to normal hyperventilation during heat exposure. Its occurrence is correlated with the rate of rise of blood pH and usually follows short periods of very intense heat stress. The serum calcium level is normal.

Heat syncope

Heat syncope usually occurs in unacclimatized individuals during the early stages of heat exposure. It is particularly common in the elderly. When cutaneous vessels dilate to enhance heat delivery to the skin, peripheral venous pooling may occur and precipitate hypotension. The mechanism is therefore similar to other forms of postural syncope and is self-limiting when the patient is supine. Treatment consists simply of rest and removal from the heat stress.

Prickly heat

Prickly heat is an acute inflammatory disorder of the skin caused primarily by blockage of sweat pores and most commonly occurs in the hot humid tropics. Maceration of the stratum corneum ensues and infection especially by staphylococci develops. In the acute phase a superficial vesicular eruption occurs associated with intense itching and often confined to clothed areas. Deeper obstruction of the sweat ducts may develop which rupture producing dermal vesicles. This is known as the profunda stage which may last for weeks. Profunda vesicles are not pruritic. The disorder can be avoided by wearing loose light clean clothing and avoiding situations provoking heavy continuous sweating.

Chlorhexidine in a light cream or lotion is the agent of choice for local treatment. Salicylic acid 1 per cent may be added to assist desquamation. Oral antibiotics may be considered in severe cases.

Anhidrotic heat exhaustion

Severe or recurrent prickly heat can occasionally compromise sweat loss and impair temperature regulation. Strenuous exercise under these circumstances can precipitate anhidrotic heat exhaustion which is seen particularly in military personnel exercising in the tropics.

Heat exhaustion

Heat exhaustion is characterized by the non-specific systemic

mic symptoms of weakness, fatigue, headache, dizziness, anorexia, nausea, vomiting, and occasionally muscle cramps. Orthostatic hypotension, tachycardia, and syncope may occur. Heat exhaustion may be differentiated from heat stroke by modest elevation in core temperature (usually less than 40°C) and absence of severe central nervous system disturbance. Slight confusion and irritability can, however, be present. The marked elevation of hepatic transaminases invariably seen in heat stroke does not occur.

Box 5.1 Heat illness syndromes

Minor
- Heat oedema
- Heat cramps
- Heat syncope
- Prickly heat
- Anhidrotic heat exhaustion
- Heat exhaustion

Major
- Heat stroke
  — Classical
  — Exertional

Heat exhaustion can result from water depletion or salt depletion. Water depletion heat exhaustion occurs following inadequate fluid replacement during exercise in hot climates. Inadequate fluid intake results in progressive dehydration and hypernatraemia. Salt depletion heat exhaustion occurs when large sweat losses are adequately replaced by water but insufficient salt and therefore hyponatraemia develops. Water depletion heat exhaustion develops rapidly over hours whereas heat exhaustion due to salt depletion develops over several days.

Pure forms of heat exhaustion are uncommon and most cases involve a mixed salt and water depletion.

Treatment consists of rest in a cool environment where spontaneous cooling should occur. Fluid and electrolyte replacement should be guided by urea and electrolyte measurements and laboratory and clinical assessment of the level of hydration. In mild cases, oral rehydration with 0.1

per cent saline solution may suffice. In more severe cases, volume replacement may be achieved by an infusion of normal saline until the patient is haemodynamically stable. Following volume restoration, body water deficits should be replaced over at least 48 hours. Rapid correction may precipitate cerebral oedema and convulsions.

Heat stroke

- **Classical heat stroke   Exertional heat stroke
   Differential diagnosis**

In heat stroke, the homeostatic thermoregulatory mechanisms are unable to cope with the demands of the heat stress. The body temperature rises precipitously resulting in multisystem tissue damage and organ dysfunction. The risk of permanent damage or death is directly related to the length of time the patient's temperature is elevated.

Heat stroke is characterized by hyperthermia with temperatures usually greater than 41 °C and severe central nervous system disturbance. Sweating may be present or absent.

It is useful to subdivide heat stroke victims into two groups associated with significantly different presentations.

Classical heat stroke occurs in epidemics during heat waves and primarily affects the elderly, the debilitated, and infants. It tends to develop over a few days and many patients are significantly dehydrated and have ceased sweating. Patients are often on diuretics or anticholinergic medication.

Exertional heat stroke occurs in young healthy individuals, such as athletes and military recruits, and often when they are unacclimatized. The onset is rapid, occurring within a few hours and usually there is insufficient time for dehydration to develop. Sweating is present in about 50 per cent of cases.

Prodromal symptoms and signs may occur. The victim may notice a decrease in the ability to concentrate, feel increasingly hot, and experience decreased sweating. Individuals who continue to exercise develop chills, throbbing headache, nausea, dizziness, and ataxia, and particularly piloerection on the chest and upper arms.

Heat stroke may be confused with other disorders particularly where fever is accompanied by central nervous system disturbance (Box 5.2). Heat stroke is a diagnosis of exclusion but cooling should be initiated immediately.

Box 5.2 Differential diagnosis of heat stroke

- Meningitis
- Encephalitis
- Intracranial haemorrhage
- Thyrotoxic crisis
- Drug-induced hyperthermic syndromes
- Delirium tremens
- Malaria
- Heat exhaustion

Clinical features of heat stroke

- CNS dysfunction   Hypotension   Bleeding diathesis   Hepatic damage   Rhabdomyolysis   Acute renal failure   Acid–base and electrolyte disturbance

CNS dysfunction

Signs of profound central nervous system disturbance dominate the early presentation of heat stroke. Delirium and coma are characteristic. Convulsions are common and often precipitated by therapeutic intervention. Oculogyric crisis, coarse tremor, and muscle rigidity may mimic seizures. The pupils may be fixed and dilated and decerebrate posturing and hemiplegias can occur. All changes are potentially reversible although permanent sequelae can occur with the cerebellum being particularly at risk.

Hypotension

Patients with heat stroke are often hypotensive. The reduction in total peripheral resistance from cutaneous vasodilation leads to a high output state manifested by a tachycardia and wide pulse pressure. As the core temperature continues to rise this compensation fails and the blood pressure falls. Occasionally hyperthermia may induce myocardial injury and necrosis and a decrease in ventricular contractility. This hypodynamic state caused by cardiogenic shock may be compounded by hypovolaemia. Arrhythmias in heat stroke may occur because of primary myocardial injury, acidosis, hyperkalaemia or hypokalaemia, and hypoxia.

Bleeding diathesis

A bleeding tendency is common in severe heat stroke and is a poor prognostic sign. The aetiology of the coagulation disturbance is complex. Thermal injury to endothelium releases thromboplastins which result in intravascular thrombosis and secondary fibrinolysis. Progression to disseminated intravascular coagulation can develop. Direct damage to platelets and to bone marrow causing thrombocytopenia can occur. In addition, hepatic dysfunction impairs production of clotting factors.

Hepatic damage

Hepatic damage is a consistent feature of heat stroke and is confirmed by elevated levels of aspartate aminotransferase (AST), alanine aminotransferase (ALT), and lactate dehydrogenase (LDH). Levels of AST greater than 1000 units are a poor prognostic factor. Jaundice typically occurs 24–72 hours after the onset of heat stroke and gradually resolves in survivors with no permanent impairment of liver function.

Rhabdomyolysis

Skeletal muscles show widespread degeneration of fibres. Rhabdomyolysis releases myoglobin, potassium, creatinine phosphokinase, and purines (which are metabolized to uric acid) into the circulation.

Acute renal failure

Acute oliguric renal failure may develop in 25–30 per cent of exertional heat stroke victims and 5 per cent of classic heat stroke cases (Box 5.3).

> **Box 5.3 Factors contributing to acute renal failure**
> - Ischaemia
> - Myoglobinuria
> - Hyperuricaemia
> - Disseminated intravascular coagulation
> - Direct thermal injury

Acid–base and electrolyte disturbance

Hyperventilation and resulting respiratory alkalosis is a physiological response to heat stress, but may be severe enough to produce tetany. Although most patients with classical heat stroke present with respiratory alkalosis, a lactic acidosis is commonly seen in cases of exertional heat stroke. When a metabolic acidosis occurs in patients with classical heat stroke it reflects severe tissue hypoperfusion and is associated with a poor prognosis.

Both hypokalaemia and hyperkalaemia may occur in heat stroke. Patients frequently have a pre-existing potassium deficit following stimulation of aldosterone during the acclimatization process. Frank hypokalaemia is common in patients with classic heat stroke. Hyperkalaemia may occur in exertional heat stroke following rhabdomyolysis or acute renal failure and can in some circumstances precipitate cardiac arrest.

Hypoglycaemia is common in exertional heat stroke as a result of intense muscle glucose utilization and hepatic damage which results in impaired gluconeogenesis.

Management of heat stroke

- **At the scene** Immediate treatment Methods of cooling Additional measures

At the scene

Cooling should begin in the prehospital phase. Mortality is primarily related to the duration of hyperthermia and therefore minutes count. The patient should be removed to a cool place and as much clothing removed as is practicable. Water can be splashed on the patient and evaporation assisted by fanning. Commercial chemical cold packs or ice water soaks can be applied to the axillae, neck, groins, and scalp. Evacuation to hospital should not be delayed by these measures which can be continued en route.

Immediate treatment

The diagnosis of heat stroke should be considered in all patients presenting to the Accident and Emergency department with signs of major central nervous system dysfunction developing in hot conditions.

The airway should be protected because coma, convulsions, and vomiting are common.

Assisted ventilation may be needed. High flow oxygen is required to support the hypermetabolic state of the tissues.

Peripheral intravenous access should be established and blood drawn for full blood count, urea and electrolytes, glucose, creatinine, ALT, AST, LDH, uric and lactic acid, calcium, and a coagulation screen. A reagent strip glucose determination should be performed and 50 ml of 50 per cent dextrose administered intravenously if indicated. Arterial blood gas determinations are required. Look for malaria if this is a potential hazard.

A rectal thermometer should be placed to obtain continuous core temperature readings. Cooling should be started immediately if the temperature is above 41°C after resuscitative measures have been instituted.

Methods of cooling

The methods employed for cooling depend on the resources and experience available. Ice water bath immersion depends on the direct conduction of heat from the body to the cold water which is 25 times greater than the conduction of air at the same temperature. It is difficult to manage comatose or convulsing patients and to maintain adequate monitoring while the patient is in the water bath. Intense peripheral constriction may shunt blood away from the skin and induce skin temperatures of 28°C or lower which may provoke shivering and endogenous heat generation. It is also extremely uncomfortable for the patient.

Evaporative measures of cooling have been shown to be as effective as ice water immersion and do not require sophisticated equipment. Evaporation of 1 gram of water removes 590 calories of heat compared to an 80 calorie loss on melting 1 gram of ice. Maximal vaporization depends on a high skin-to-air temperature gradient, adequate cutaneous blood flow, rapid air flow, and low relative humidity in the treatment area.

A practical way to keep the skin moist is via a mist sprayer. The water should not be cold but approximately 15°C. Effective circulation of room air can be achieved by using portable electric fans. The patient should always be completely naked and not covered with a sheet or towel. Ideally, the patient should be suspended on a net surface but this is not often practicable.

In Accident and Emergency departments which do not routinely handle patients with hyperthermia, the best approach is a combination of evaporative methods and direct application of ice water soaks to the axillae, groins, neck, and scalp. The soaks should be changed frequently.

An adequate rate of cooling is 0.1°C per minute. An attempt should be made to correct the hyperthermia within 45–60 minutes. When the core temperature falls to 38–39°C, cooling should be stopped to avoid an hypothermic overshoot. Rebound hyperthermia may occur 3–6 hours after successful cooling and thermoregulatory instability may exist for days.

Additional measures

Hypotension is common in heat stroke and is usually due to high output cardiac failure as a result of shunts through the dilated skin vessels. The hypotension usually resolves on cooling. If severe hypotension persists after the first few degrees of cooling a bolus of 200–400 ml of normal saline can be administered. If the blood pressure rises, further replacement can be given with careful monitoring. Pulmonary oedema may be precipitated by over-zealous fluid administration after cooling when cutaneous vasoconstriction returns large volumes of fluid to the central circulation. A central venous pressure line may be of value but in severe hypotension and refractory cases a Swan Ganz catheter is preferable (care is needed if coagulation is abnormal).

Electrocardiographic monitoring is required. A nasogastric tube should be passed to decompress the stomach and to monitor the onset of gastrointestinal bleeding. The bladder should be catheterized to monitor urine output.

Chlorpromazine 25–50 mg intravenously is usually recommended if shivering occurs during cooling. However, it can lower the seizure threshold, aggravate hypotension, and has anticholinergic properties which may interfere with sweating. It should therefore only be used if the cooling rate is inadequate in the presence of vigorous shivering.

Antipyretics including aspirin are not indicated.

Convulsions can be treated with 2.5 mg aliquots of intravenous diazepam.

Peritoneal dialysis and cardiopulmonary bypass may be required in those severe cases not responding to the above measures.

Prevention

- Avoidance measures   Fluid intake   Measurement of heat stress

Avoidance measures

Heat illness is associated with considerable individual susceptibility but measures can be taken to minimize the risk particularly when exercising in hot climates.

Strenuous physical activity should be avoided if a risk factor such as febrile illness or diarrhoea is present and timed to avoid intense sunlight exposure and the hottest times of the day.

Light-coloured loose-fitting clothing reflects light and permits air convection and more efficient evaporative heat loss from the skin.

Fluid intake

Adequate fluid intake is crucial and should be encouraged before, during, and after exertion: 500 ml of fluid should be consumed prior to exercise and a minimum of 200–300 ml at 20 minute intervals during exertion.

Athletes are unable to monitor accurately their own fluid requirements and consumption. Measuring daily body weight while training in hot climates is safer. A weight loss of 5 per cent is a contraindication to further exercise until adequately rehydrated.

The fluid chosen for replacement should be palatable to encourage adequate intake and ideally should stimulate gastric emptying so that it rapidly enters the small bowel where absorption occurs. Gastric emptying is accelerated by large fluid volumes (500–600 ml) at cool temperatures of 10–15 °C. Weak electrolyte solutions with an osmolality of about 200 mmol/l are adequate. Solutions with higher osmolality delay gastric emptying and some commercially available preparations contain excessive and unnecessary glucose.

Supplemental salt tablets are not necessary and when taken with inadequate fluid replacement can precipitate hypernatraemic dehydration and cardiovascular collapse. Excessive salt intake may impair acclimatization by inhibiting aldosterone secretion.

Measurement of heat stress

The wet bulb globe temperature (WBGT) index is the most accurate measurement of environmental heat stress. The WBGT measures three forms of heat load: (1) a regular thermometer measures dry air temperature; (2) a wet bulb thermometer measures the effect of humidity on temperature; and (3) a globe thermometer measures the effect of radiant heat.

WBGT = 0.2 globe temperature + 0.1 dry bulb temperature + 0.7 wet bulb temperature. Ninety per cent of cases of heat stroke occur at WBGT greater than 30 °C.

Those responsible for organizing athletic activities and military training exercises should be aware of these measurements of environmental heat stress and must delay or cancel activities if the risk of heat stroke is high.

Further reading

Anonymous (1982). Management of heat stroke. *Lancet*, **2**, 910–11.

Khoegli, M. and Weiner, J. S. (1980). Heat stroke: report on 18 cases. *Lancet*, **2**, 276–8.

Shibolet, S., Lancaster, M. C., and Danon, Y. (1976). Heat stroke: A review. *Aviation Space and Environmental Medicine*, **47**, 280–301.

Whitworth, J. A. G. and Wolfman, M. J. (1983). Fatal heat stroke in a long distance runner. *British Medical Journal*, **287**, 948.

Yaqub, B. A., Al-Harthi, S. S., Al-Orainey, I. O., Laajam, M. A., and Obeid, M. T. (1986). Heat stroke at the Mekkah Pilgrimage: Clinical characteristics and course of 30 patients. *Quarterly Journal of Medicine*, **229**, 523–30.

CHAPTER 6

High-altitude illness

General introduction	91
High-altitude environment	92
Response to altitude	93
Acute altitude illness	96
Acute mountain sickness	96
High-altitude pulmonary oedema	96
High-altitude cerebral oedema	97
Retinal haemorrhages	97
Treatment	98
Prevention	99
Injury from ultraviolet radiation	101
Effects of high altitude on pre-existing medical conditions	103
Further reading	103

Key points in high altitude illness

1 Acute high-altitude illness includes the relatively benign syndrome of acute mountain sickness and the less frequent life-threatening conditions of high-altitude pulmonary oedema and cerebral oedema.
2 The incidence, severity, and duration of altitude illness vary with the rate and height of ascent as well as individual susceptibility.
3 The symptoms of acute altitude illness are caused by pathological responses to environmental hypoxia and usually develop after several hours at high altitude.
4 Gradual ascent is the safest method of prevention.
5 Attention to adequate hydration is essential at high altitude.
6 Acetazolamide can help in the prophylaxis of acute mountain sickness but has side-effects which may not be tolerated.
7 Oxygen helps most patients with acute altitude illness but relief is often only partial and deterioration may continue.
8 Immediate descent is the only safe management for victims with serious symptoms of acute mountain sickness and those with high-altitude pulmonary or cerebral oedema.

General introduction

- **Mountaineering   Trekking**

High-altitude illness represents a continuum of disorders ranging from the relatively benign syndrome of *acute mountain sickness* to the life-threatening conditions of *high-altitude pulmonary oedema* and *high-altitude cerebral oedema*. The incidence, severity, and duration of illness are highly dependent on the rate of ascent and altitude attained. Mild symptoms are experienced by 50 per cent of people ascending above 3500 metres and nearly 5 per cent of individuals experience severe symptoms at this altitude.

Increasingly, doctors are confronted by those embarking on expeditions with questions about prevention and treatment of high-altitude medical problems and the effects of altitude on pre-existing medical conditions. Doctors may also be asked to join trekking and mountaineering expeditions.

Acute altitude illness is seen in mountaineers and skiers but most frequently in trekkers. The recent popularity of trekking in the Himalayas has led to an influx of many novices to high altitudes—over 20 000 people from the United Kingdom alone travel to Nepal each year. A study of trekkers in the Thorong Pass (5400 metres) found that two-thirds were suffering from symptoms of high-altitude illness. Travel has also become much faster and easier enabling the unacclimatized traveller to go from sea level to potentially dangerous altitudes in less than a day.

Some mountaineers are well informed about altitude illness and carry emergency equipment, spend time acclimatizing, and take appropriate action when illness supervenes. Incapacitating illness, difficult terrain or bad weather may, however, make descent to safety impossible. Mountaineers who acclimatize at lower altitudes (4000–5000 metres) are still at risk from sudden onset of illness following ascent to greater heights. In a study of British expeditions to peaks over 7000 metres between 1968 and 1988, 23 deaths occurred in 533 climbers. Although most deaths were caused by accidents, disorientation, and misjudgement resulting from the effects of high altitude probably contributed significantly.

High-altitude environment

- **Barometric pressure Hypoxia Temperature Solar radiation Ultraviolet radiation Humidity**

A variety of environmental stresses accompany exposure to high altitudes. The relative amount of oxygen in the ambient air is the same at high altitude as at sea level (21 per cent). However, the ambient pressure falls with increasing altitude and so the partial pressure of oxygen also decreases (Fig. 6.1). At sea level the average barometric pressure is 101.3 kPa. At 5500 metres the barometric pressure is half of that at sea level, and at the summit of Mount Everest (8848 metres)

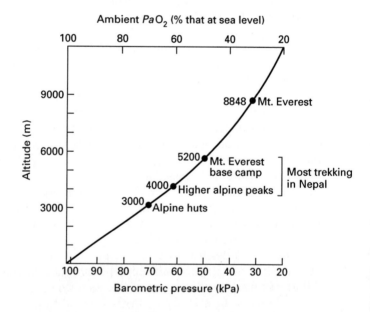

Fig. 6.1 • Relation between altitude, barometric pressure, and partial pressure of oxygen in ambient air as a percentage of that at sea level. [Adapted from Frisancho, A. R. (1975). *Science*, **187**, 314 © AAAS.]

the partial pressure of oxygen falls to about one-third. The symptoms of altitude induced illness are caused by the pathophysiological response to this environmental hypoxia.

Temperature falls with increasing altitude by about 1 °C for every 150 metres and severe cold stresses can result at high altitude. The average temperature on the summit of Mount Everest is −40°C, although there is a wide seasonal and daily variation. The effect of such low temperatures at altitude is exacerbated by high winds which result in extremely severe wind chill-effects (Chapter 2, p. 31).

The intensity of solar radiation increases at altitude due to decreased filtration by the atmosphere and reflection from snow. Despite a reduction in temperature at altitude extreme variation in environmental temperature occurs, due to fluctuations in the degree of solar radiation.

At high altitude there is less filtering of sunlight, particularly the ultraviolet (UV) wavelengths. Excessive UV radiation, exacerbated by reflectance from snow, can cause sunburn and snow-blindness.

Humidity is extremely low at high altitude. This exacerbates insensible fluid losses which are high from increased ventilation and sweating during exercise. Climbers may require a fluid intake of 3–4 litres per day to maintain an adequate urine output at altitudes of over 6000 metres.

Response to altitude

- **Hypoxic ventilatory response   Periodic breathing
Fluid retention and redistribution**

Although the underlying cause of acute mountain sickness is hypoxia associated with high altitude ascent, the syndrome which develops is different from that of acute hypoxia. A time lag in onset of symptoms and incomplete reversal of all symptoms with oxygen, suggests that acute mountain sickness is secondary to the body's *response* to the hypoxic insult (Fig. 6.2).

An increase in ventilation is a fundamental response during acclimatization at altitude. A decrease in arterial

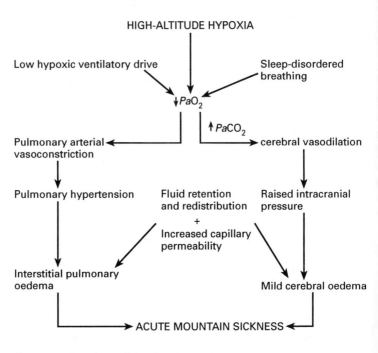

Fig. 6.2 • Pathophysiology of acute mountain sickness.

PaO_2 is detected by the carotid body, causing reflex stimulation of the medullary respiratory centre. This *hypoxic ventilatory response* begins at altitudes as low as 1500 metres and occurs within a few hours of altitude exposure. An increase in ventilation reduces the oxygen gradient between inspired and alveolar air thereby enhancing delivery.

As ventilation increases the development of hypocapnia and resulting alkalosis limits a further increase. However, within 24–48 hours renal excretion of bicarbonate restores pH and ventilation thereafter continues to increase reaching a maximum in 4–7 days at the same altitude. Further increases in ventilation and reduction in plasma bicarbonate occur at increased elevations.

People with a low hypoxic ventilatory response are particularly likely to develop acute mountain sickness. Hypoxia and elevation in the partial pressure of carbon dioxide result in increased cerebral blood flow. Hypoxia also increases capillary permeability and combined with raised pressures in the pulmonary and cerebral vessels will contribute to the development of oedema. The oedema fluid found in high altitude pulmonary oedema has the characteristics of a permeability leak rather than that of a hydrostatic leak.

A feature of sleep at altitude is *periodic breathing*. Periodic breathing is characterized by periods of hyperpnoea followed by long periods of apnoea which can result in severe hypoxaemia. Sleep imposes an additional burden on gas exchange and can precipitate acute altitude illness.

Hypoxia causes alterations in fluid and electrolyte balance. There is water retention and shifts of fluid from intracellular to extracellular compartments. The mechanisms which produce such changes are not fully understood.

A mild diuresis and contraction of plasma volume are normal adaptive responses at high altitude. The diuresis may be under the control of atrial natriuretic peptide. Some people become oliguric during their first few hours at altitude and may be prone to acute mountain sickness as their plasma volume expands. Vigorous exercise during ascent or on arrival at high altitude also results in an expansion of the plasma volume via the renin-aldosterone mechanism.

The same factors which contribute to the development of acute mountain sickness may become more pronounced and cause the symptoms of high-altitude pulmonary oedema and high-altitude cerebral oedema. However, the progression from acute mountain sickness to more serious illness in susceptible individuals must involve other mechanisms.

Acute exposure to high altitude results in increased platelet count and adhesiveness and elevated levels of other coagulation factors. This is counteracted by enhanced fibrinolytic activity. An imbalance may result in disturbed coagulation and microthrombi formation which has been implicated in high-altitude pulmonary and cerebral oedema. Occasionally at extreme altitudes, deep vein thrombosis and major pulmonary and cerebral thrombosis can occur. Dehydration at altitude will predispose to thrombosis. Women taking the contraceptive pill are also at risk.

Acute altitude illness

- Acute mountain sickness   High-altitude pulmonary oedema   High-altitude cerebral oedema   Retinal haemorrhages

Acute mountain sickness

Rapid ascent to altitudes above 3500 metres may precipitate the relatively benign syndrome of acute mountain sickness. Symptoms develop after a delay of several hours but usually within the first 36 hours. Headache, nausea, anorexia, dizziness, irritability, and insomnia are typical features. The headache is characteristically throbbing, frontal, worse in the morning and in the supine position, and exacerbated by strenuous exercise. There may be fluid retention with ankle and facial oedema.

Specific physical findings are relatively few. Localized crepitations may be found on auscultation of the chest. Fundoscopy may reveal engorged retinal vessels and retinal haemorrhages (p. 97).

High-altitude pulmonary oedema

High-altitude pulmonary oedema is usually associated with rapid ascent from sea level to high altitudes between 4000 and 5000 metres and thereafter engaging in strenuous physical exercise too quickly. Symptoms of acute mountain sickness frequently precede the condition which usually develops a day or more after ascent. Both dyspnoea and cough on exertion are common at high altitude and difficult to attribute to developing illness. However, *dyspnoea at rest* is particularly ominous and is often accompanied by a tachycardia. Chest pain and a persistent dry cough are common complaints. The condition typically worsens at night and orthopnoea is common. Progressive respiratory distress is accompanied by cyanosis and expectoration of blood-tinged frothy sputum. Widespread crepitations are heard on auscultation. Deterioration may occur over a few hours or occasionally much more rapidly in a matter of minutes.

Chest X-ray taken after evacuation to hospital typically reveals infiltrates which are fluffy, patchy, and frequently asymmetrically distributed. Pulmonary vessels may be prominent. Appearances are often more striking than would be expected from the clinical findings.

High-altitude cerebral oedema

High-altitude cerebral oedema is the most severe but least common form of altitude illness. Some people, usually with symptoms of acute mountain sickness and particularly headache, suddenly deteriorate developing prominent truncal ataxia, confusion, and altered consciousness. Victims typically walk unsteadily with a broad-based gait. Auditory and visual hallucinations are common and often very vivid. Double vision and cranial nerve palsies may occur. Neck stiffness, hyperreflexia, and extensor plantar responses may be found. Fits are rare. Examination of the retinae may show papilloedema or retinal haemorrhages. A degree of pulmonary oedema may be evident.

Retinal haemorrhages

Retinal haemorrhages can occur at altitudes above 4000 metres, are more often seen in people going to high altitude for the first time and usually appear during the first few days after ascent. The haemorrhages are often multiple, flame-shaped, and adjacent to retinal vessels which are usually engorged. The condition is almost always symptomless and self-limiting although blurring of vision can occur if the haemorrhages are sited near the macula. They may result from microvascular rupture of dilated vessels when intraocular pressure rises, for example during coughing which is common at high altitude. Such haemorrhages are often seen in patients with acute mountain sickness but there is no strong correlation with severity and they are not a sign of impending cerebral oedema. This is in distinction to the haemorrhages seen with papilloedema which are associated with high altitude cerebral oedema.

98 • High-altitude illness

Treatment

- Descent   Oxygen   Pressurization bag   Medication

Acute mountain sickness usually resolves spontaneously if ascent is halted. Paracetamol for headache and an antiemetic may be required. Within a few days most victims recover and are able to gradually resume activity including further ascent if desired.

Descent

Prompt recognition of deteriorating acute mountain sickness and the early symptoms of high-altitude pulmonary and cerebral oedema is vital. *The only safe management for a victim is immediate descent.* There is no definite safe level, although evacuation to below 2500 metres is advisable. Even if this cannot be achieved, marked improvement often accompanies modest reduction (300–500 metres) in altitude.

Occasionally, however, descent may be impossible due to bad weather, avalanches, or darkness. The victim may be too ill to climb down and beyond the reach of the rescue services. In these circumstances emergency treatment may improve the victim's condition and facilitate descent.

Oxygen

If available, oxygen should be given at 3–4 litres per minute by a tight-fitting mask. Although most patients benefit from oxygen, relief is often only partial and deterioration may continue. Continuous flow oxygen systems using lightweight oxygen bottles (3–7 kg) are the most common type carried on expeditions. Supplies may be conserved if necessary by giving lower flow rates of 0.5–1 litre per minute, particularly at night. Oxygen is not a substitute for and should not delay descent.

Pressurization bag

Recently, high altitude illness has been treated by acute pressurization in lightweight (less than 5 kg) fabric hyperbaric chambers. Pressurization is achieved by a hand or foot pump. A high air flow is required to keep the concentration of CO_2 inside the bag below 1 per cent. The increase in pressure in the bag above ambient pressure can simulate a descent of 1500–2500 metres depending on the altitude and type of bag used. Alleviation of symptoms may facilitate descent and may be as effective as oxygen therapy.

Medication

Dexamethasone can significantly reduce symptoms in patients with established and disabling acute mountain sickness and high-altitude pulmonary and cerebral oedema. An initial dose of 8 mg given orally or intramuscularly should be followed by 4 mg every 6 hours. The dose should be reduced gradually and stopped after 7–10 days.

Frusemide (40–80 mg) or bumetanide (1–2 mg) given orally or intravenously may induce a diuresis and have been used in patients with pulmonary and cerebral oedema. However, these loop diuretics do not seem to be as effective in high-altitude pulmonary oedema as they are in cardiogenic pulmonary oedema. Their use may, in fact, exacerbate the illness by increasing intracellular dehydration and precipitating thrombosis secondary to haemoconcentration.

Nifedipine (20 mg slow release 6 hourly) may be of value in high-altitude pulmonary oedema. This calcium channel antagonist reduces pulmonary arterial pressure and therefore facilitates the resorption of oedema fluid. Headache is a recognized side-effect.

Acetazolamide (see below) is not of value in established high-altitude pulmonary and cerebral oedema. It may help in acute mountain sickness when severe headache (particularly nocturnal) is an isolated complaint.

Prevention

- **Controlled ascent   Acetazolamide   Dexamethasone**

Controlled ascent is the safest method of prevention. Current recommendations are to avoid travelling from sea level to sleeping altitudes of greater than 3000 metres on day one and to spend two to three nights at this altitude before going higher. Over-exertion should be avoided for the first few days. Beyond 3000 metres, climbers should not sleep higher than 300 metres above their previous camp. Every 2 to 3 days a rest day is advisable. However, such guidelines are not popular and not often adhered to.

Adequate hydration is essential at high altitude and a fluid intake of 3 litres per day is recommended.

Acetazolamide is the drug of choice for prophylaxis of acute mountain sickness. However, it is not a substitute for acclimatization and by reducing minor symptoms may encourage some people to ascend too quickly and develop more serious illness without warning.

During ascent to high altitude renal reabsorption of bicarbonate partially corrects the respiratory alkalosis and initially limits the ventilatory response to hypoxia. Acetazolamide is a carbonic anhydrase inhibitor and slows the hydration of carbon dioxide.

$$CO_2 + H_2O \underset{\rightleftharpoons}{CA} H_2CO_3 \underset{\rightleftharpoons}{CA} H^+ + HCO_3$$

In the kidney, bicarbonate reabsorption is reduced and metabolic acidosis results. The acidosis stimulates ventilation and arterial PaO_2 increases. Prophylaxis with acetazolamide mimics the acclimatized state of acid–base balance such that during the first day of altitude exposure, minute ventilation and arterial blood gas values are typical of those not usually observed until day five in untreated individuals.

Acetazolamide 500 mg daily as a sustained release preparation should be commenced 3 days before ascending to 3000 metres and continued for 2–3 days at altitude until acclimatization is established.

Side-effects of acetazolamide occur and may not be tolerated. Paraesthesiae of the hands and feet, gastrointestinal upset, and drowsiness may occur. Carbonated beverages may taste flat as acetazolamide inhibits the instant hydration of carbon dioxide on the tongue. A trial of acetazolamide at sea level for 2 days several weeks before ascending to altitude will indicate the potential for individual side-effects. Sensitivity to sulphonamides is a contraindication.

Dexamethasone is not recommended for routine prophylaxis but may be of value in those embarking on urgent rescue operations especially at very high altitudes.

The use of sedatives to improve the quality of sleep at altitude should be avoided as they depress nocturnal ventilation. It is recommended that the oral contraceptive pill should not be taken because of the associated fluid retention and risk of thrombosis.

An episode of acute altitude illness is not a contraindication to subsequent altitude exposure but appropriate prophylaxis and education to ensure recognition of early symptoms are essential.

Injury from ultraviolent radiation

- **Sunburn   Snow-blindness**

The electromagnetic radiation emitted from the sun covers a wide spectrum of wavelengths and includes ultraviolet (UV) radiation between 200–400 nm. This UV band is made up of UVA (320–400 nm), UVB (290–320 nm), and UVC (200–290 nm). UVC is filtered out by the ozone layer. Further absorption of UV radiation occurs as it passes through the atmosphere, UVA to a greater extent than UVB. Both UVA and UVB radiation penetrate clouds.

Sunburn

Sunburn results primarily from UVB exposure although the injury is potentiated by UVA radiation. At high altitude there is less filtering of UV radiation by the atmosphere. The intensity of UV radiation is therefore significantly increased and there is a disproportionate increase in intensity of the more harmful UVB spectrum. Reflection of a high proportion of UV radiation from snow (albedo) may lead to severe sunburn in areas not directly exposed such as the under surface of the chin and nose.

Excessive exposure of the skin to UV radiation will produce erythema after a delay of several hours. Blistering may develop in severe cases reflecting damage to proteins and nucleic acids in the skin.

Clothing and an efficient sunscreen can prevent sunburn. p-Aminobenzoic acid (PABA) and its esters are efficient sunscreens but only filter and absorb the UVB spectrum. Non-PABA sunscreens (e.g. oxybenzone, cinnamates) are commonly used in conjunction with PABA sunscreens to enhance the spectrum of protection to include some of the UVA spectrum. Physical blocking agents, such as zinc oxide and titanium dioxide, result in an opaque layer on application and exert their sunscreen effect by reflecting and scattering both UVA and UVB radiation. They are less cosmetically acceptable compared to other sunscreens and generally only used on the lips, nose, and ears. Acute contact dermatitis can occasionally occur with PABA preparations but are rare with non-PABA preparations. PABA may stain clothes yellow but its esters do not.

The sun protection factor (SPF) quoted for sunscreens only applies to UVB screening and does not apply to UVA sunscreens. SPF measures the factor by which sun exposure can be increased before a minimal degree of erythema occurs by comparison with unprotected skin. It should, however, be noted that the SPF quoted often over-estimates the UV radiation protection provided in actual outdoor use by 20–30 per cent. It is recommended that a combined sunscreen with both UVB and UVA protection is used at high altitude and with a SPF of 15–20 or higher. The lips should also be protected by a lipsalve with sunscreen. Sunscreens have to be applied regularly to provide the appropriate level of protection.

Snow-blindness

Exposure of the corneal epithelium to excessive UV radiation causes keratitis. There is no initial sensation during exposure but after a delay of a few hours a gritty feeling occurs in the eyes. Intense injection of the conjunctivae and profuse lacrimation is seen. There may be severe photophobia and blepharospasm. Corneal pitting (punctate keratosis) will be identified after fluorescein staining in serious cases.

The condition may be prevented by wearing appropriate protective sunglasses with side protectors. The pain and spasm can be relieved by instilling anaesthetic eye drops. In cases of corneal ulceration antibiotic ointment should be applied and an eye patch worn for 24 hours.

Effect of high altitude on pre-existing medical conditions

Advice may be sought by those proposing to travel to high altitude about the effects on pre-existing medical conditions. Generally, individuals should be as fit as possible before embarking on a trekking or climbing expedition.

Although an attack of asthma may be precipitated by the cold air and exercise at altitude, absence of allergens usually results in less trouble than at sea level. Inhalers may be a problem at high altitude and extremes of temperature; dry powder inhalers may be better than pressurized aerosols. Energy requirements during exercise at altitude are very variable and careful control of blood glucose is required by diabetics. Those with mild or controlled hypertension may go to high altitude. There is no evidence that epiletic fits are more frequent. High altitude will worsen any symptoms associated with cardiovascular disease or impaired lung function. Sickle-cell crisis can be precipitated in those with the trait. The risks of travelling to altitude during pregnancy are unknown but it is recommended that ascent is restricted to modest altitudes.

Further reading

Bartsch, P., Merki, B., Hofstetter, D., Maggiorini, M., Kayser, B., and Oelz, O. (1993). Treatment of acute mountain sickness by simulated descent; a randomised controlled trial. *British Medical Journal*, **306**, 1098–101.

Dickinson, J. G. (1987. Acetazolamide in acute mountain sickness. *British Medical Journal*, **295**, 1161–2.

Hacket, P. H. (1980). *Mountain sickness; prevention, recognition and treatment*. American Alpine Club, New York.

Heath, D. and Williams, D. R. (1989). *High altitude medicine and pathology* (3rd edn). Butterworths, London.

Johnson, T. S. and Rock, P. B. (1988). Acute mountain sickness. *New England Journal of Medicine*, **319**, 841–5.

Pollard, A. J. (1992). Altitude induced illness. *British Medical Journal*, **304**, 1324–5.

Ward, M. P., Milledge, J. S., and West, J. B. (1989). *High altitude medicine and physiology*. Chapman and Hall, London.

CHAPTER 7

Lightning injuries

General introduction	107
The lightning stroke	108
Mechanism of injury	110
Clinical points	111
Avoiding lightning injuries	113
Further reading	114

Key points in lightning injuries

1 Lightning may cause cardiac arrest. The heart often restarts spontaneously but respiratory arrest may then cause hypoxia and secondary cardiac arrest which is fatal without immediate resuscitation.
2 The flashover effect generally results in victims sustaining superficial burns only. Deep muscle damage and myoglobinuria are rare in lightning injury.
3 Coma, fits, transient paralysis, confusion and amnesia, ocular injury, and rupture of the tympanic membrane are typical clinical features.
4 Injuries may occur when the victim is thrown to the ground.
5 Arborescent or fern-like skin markings (Lichtenberg figures) are pathognomonic of lightning injury.
6 Sensible measures can be taken to avoid the risk of lightning injury. Weather forecasts of thunderstorms should be respected.

General introduction

- Incidence   Predisposing factors

Lightning kills on average four people a year in the United Kingdom. For every person killed by a direct lightning strike four or more are injured. Men are six times more likely to be killed as women. Thirty per cent of victims are killed in groups of two or more.

In the United States, lightning kills more people every year (about 150–300 deaths currently) than any other naturally occurring disaster including hurricanes, earthquakes, and floods. Lightning strikes are also much more common in tropical countries.

Although lightning may occur without rain it cannot without thunder. There are between 1500 and 2000 thunderstorms throughout the world at any moment with lightning striking about 6000 times every minute. Lightning strikes are prevalent in high mountainous areas and around large areas of water.

The frequency of injury and death from lightning strikes is largely determined by geographical factors but population density and behaviour both play a part. Lightning tends to seek out the highest object. Tall buildings and structures are frequently struck but afford protection to town and city dwellers. Consequently lightning injuries occur more commonly in rural than urban areas. Farmers and other workers in isolated areas used to be the group most frequently injured by lightning strikes. However, with the gravitation of populations to urban areas, hill climbers, golfers, and those engaged in other outdoor recreational activities seem to be at greatest risk.

Lightning may well strike in the same place twice. If the factors which facilitate the original strike are still *in situ* then further strikes will be more prevalent in that area.

The lightning strike

- **Lightning discharge   Conductor rods   Thunder**

When warm, moist air of a low pressure system rises the moisture condenses and a cloud develops. If convection is vigorous a cumulonimbus may develop which contains a mixture of water droplets, ice crystals, hailstones, and ice pellets (graupel). Turbulent air currents in the cloud cause these hydrometeors to collide transferring electrical charges between them. The larger ones gain a negative charge as they fall causing the lower layer of the thundercloud to become negatively charged and leaving the upper layer positively charged. The lower layer of the thundercloud induces a strong positive charge in the earth beneath. Eventually, the potential difference overcomes the insulating effect of the air between and the charge dissipates as lightning.

Lightning discharge

A lightning flash begins with a relatively weak and slow leader stroke from the cloud which leaves a trail of ionized air (Fig. 7.1). As a branch of the leader comes to within 100 metres of the ground the negative charge in the air attracts a positive charge just above the ground. One or more pilot stroke discharges propagate upwards to attach to the leader, 10–20 metres above the ground surface. When contact is made the channel of ionized air is completed and the air resistance falls. The resistance drop allows a much more rapid and highly charged return stroke to pass from the ground to the cloud. After a fractional pause, another leader propagates down the ionized channel from the preceding return stroke and precipitates a second return stroke. A typical lightning flash consists of three or four leaders and return strokes.

The tip of the leader stroke is the most luminous of the sequence of strokes in each lightning discharge as energy is expended in forming the ionized channel. This gives the impression of lightning travelling from cloud to earth despite a greater energy being dissipated during the return stroke in the opposite direction. The direction of the return stroke is

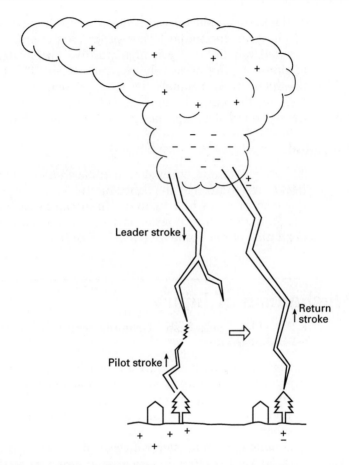

Fig. 7.1 • Lightning discharge.

not visually perceived because of its immense speed and is seen as a brightning or flickering of the ionized pathway. Sometimes the leader propagates only part of the way down the existing ionized channel and then forges a different path to the ground. This is seen as forked lightning. Sheet lightning may also occur which is a shapeless flash of light from lightning discharges within clouds.

Conductor rods

A lightning conductor rod protects a building by inducing discharge of the initial leader stroke as it approaches. The pointed rod builds up a high positive charge density in response to the negatively charged leader. The lightning conductor then channels the return stroke along its cable keeping the current safely outside the building. Isolated trees and tall structures behave in a similar fashion.

Thunder

Thunder is due to an explosive expansion of air which is heated and ionized by the lightning stroke. The approximate distance in miles to a lightning flash can be estimated by counting the time in seconds between the flash and the beginning of the thunder clap and dividing by five.

Mechanism of injury

- Direct hit   Side-flash   Ground current   Blunt trauma   Flashover effect

A person may be injured by a direct lightning strike. However, more frequently a side-flash injury occurs when the current jumps from the primary conductor, such as a tree, on to a victim. Side-flashes may also occur from one person to another.

Ground current (or step voltage) injury occurs when lightning strikes the ground close to a person. A large potential difference develops between the victim's feet setting up a current through the victim's legs and lower part of the body. Although multiple victims may result from a ground current strike, little if any current flows through the heart or brain, and death from ground current is rare. Swimmers may also be affected by this mechanism of injury.

Blunt trauma may result from the explosive force of the shock wave produced by nearby strikes or impact with the ground or other objects.

There are significant differences between injury produced

by lightning compared to other high-voltage sources. The most important factor is the very much shorter duration of exposure to the current produced by lightning. In other high-voltage injury the sustained current flow is responsible for the severe tissue damage. Exposure to lightning is so brief that there is insufficient time for significant skin burn or deep muscle damage to occur. The energy is dissipated instead by passing over the outside of the body as 'flashover effect'. This may vaporize sweat and rain water and blast off clothes and shoes.

Clinical points

- **Cardiopulmonary arrest   Diagnosis and management**

Cardiopulmonary arrest

A proportion of the current associated with a lightning strike may leak internally and act like a massive DC countershock causing an asystolic cardiac arrest. Automaticity of the healthy heart usually supervenes and cardiac contractions return spontaneously. However, the accompanying respiratory arrest from paralysis of the medullary respiratory centre is often prolonged and if ventilatory support is not promptly provided at the scene, the resulting hypoxia can cause secondary cardiac arrhythmias and cardiac arrest. Treatment of cardiac arrhythmias and cardiopulmonary arrest should follow standard treatment protocols.

Diagnosis and management

A 12 lead electrocardiogram should be performed in all victims of a lightning strike and blood sent for cardiac enzyme determinations. Although it is common to find non-specific ST–T-wave changes, established infarction patterns are rare. Cardiac enzyme levels may rise reflecting areas of focal cardiac necrosis.

Depressed conscious level on admission may be due to direct injury as the current traverses the brain, or secondary to hypoxia or closed head injury. A CT scan will help define the extent of brain injury in patients with altered conscious level. There may be evidence of coagulation of the cerebral cortex or extradural, subdural, or intraventricular haemorrhages. Fits and transient paralysis (keraunoparalysis) can occur. About two-thirds of victims have some degree of lower extremity paralysis and almost one-third have upper extremity paralysis. The weakness resembles post-ictal paralysis and generally resolves in 24–48 hours. Victims often remain confused and amnesic for several days.

Frequently, the limbs are cold, mottled, and pulseless secondary to instability of the sympathetic nervous system. The intense vascular spasm associated with this automatic instability usually resolves spontaneously over several hours.

The flashover effect generally results in victims sustaining only superficial burns. A characteristic erythematous rash in a fern-like or arborescent pattern appears in about a fifth of victims. The skin pattern, known as Lichtenberg figures, is pathognomonic of lightning injury. They are not true burns and are attributed to electron showers tracking over the skin. They typically resolve in 24 hours. Occasionally, metal such as a buckle adjacent to the skin may cause deeper burns when heated by the electrical energy.

Deep muscle damage following lightning injury is rare in comparison to other electric shocks and therefore myoglobinuria and resulting renal damage seldom occur. Fasciotomy is rarely, if ever, required.

Significant blunt trauma can be sustained by the victim thrown by the strike. Appropriate radiological investigation including a lateral cervical spine X-ray will be required in such patients. Direct blunt injury from the explosive force of the strike may occasionally occur resulting in bursting soft tissue injury and fractures which are seen typically in the feet.

Eye injuries are common. Cataracts are the most frequent injury and may present in the first few days or up to two years after the lightning injury. They are often bilateral. Corneal damage, hyphaema, uveitis, retinal detachment, vitreous haemorrhage, and optic atrophy are also seen. Dilated and unreactive pupils should never be used as a prognostic sign as autonomic dysfunction may be the primary cause.

Temporary deafness is very common and more than half the victims have tympanic membrane rupture due to the shock wave.

The diagnosis of lightning injury may be difficult if the strike is not witnessed. Other causes of depressed conscious level, convulsions, cardiac arrhythmias, and blunt trauma may have to be excluded. Patients may sometimes be considered victims of assault because of disarray of their clothing. Careful physical examination to identify the typical injury pattern associated with lightning strikes will help clarify the diagnosis.

All victims of a lightning strike should be admitted for 24 hours' observation and those with electrocardiographic changes and arrhythmias will require continuous ECG monitoring.

Avoiding lightning injury

Weather forecasts of thunderstorms should be respected. It is sensible to obtain weather reports prior to going on an expedition.

If a thunderstorm is imminent shelter should be sought in a building or if caught in the open, ditches and valleys may afford protection unless filled with water. Tents afford little protection because the metal support poles may act as lightning conductor rods.

A group of people exposed during a thunderstorm should avoid contact and remain several metres apart to decrease the number of potential victims injured by ground current or side-flashes.

Isolated trees and hilltops should be avoided. About one-quarter of victims killed by lightning were sheltering under trees.

The safest position if caught in a wide open space is to crouch with the feet close together to minimize the potential difference, with hands on the knees.

During a thunderstorm warning of a possible lightning strike may occur. The formation of a positive electrical charge spreading upwards to meet the leader strike may cause hair to stand on end or metal objects nearby to buzz or crackle. Lightning may not always follow but it is best to move away!

Metal fences, pipelines, and railway tracks can conduct lightning which has struck some distance away. Metal objects, such as umbrellas and golf clubs, should be discarded. Golfers should remove their shoes if these have metal studs.

Lightning is attracted to metal masts and other objects projecting above the surface of water and the current can pass through water to injure swimmers. Therefore those engaged in water sports should make for the shore.

When lightning strikes a vehicle the current flows around the occupant through the metal bodywork before arcing the last few centimetres from the tyres to ground. It is therefore safer to remain in the vehicle.

Lightning is also potentially dangerous to people indoors. Lightning frequently strikes aerials and the cable may conduct the current to the television set with risk of explosion and fire. Similarly, lightning injury has been recorded whilst using a telephone during a thunderstorm.

Further reading

Cooper, M. A. (1984). Electrical and lightning injuries. *Emergency Medicine Clinics of North America*, **2**, 489–501.

Elsom, D. (1989). Learn to live with lightning. *New Scientist*, **122** (No. 1670), 54–8.

Hanson, N. C. and McIlwraith, G. R. (1973). Lightning injury; two case histories and a review of management. *British Medical Journal*, **4**, 271–4.

Smith, T. (1991). On lightning (Editorial). *British Medical Journal*, **303**, 1563.

CHAPTER 8

Chemical accidents and emergencies

General introduction	117
At the scene	118
Chemical contamination	121
Toxic inhalations	126
Injuries due to chemical weapons	135
Further reading	138

Key points in chemical accidents and emergencies

1. At the scene of a major chemical accident the fire service are responsible for identifying and containing the hazard. Rescue attempts should not be made until the incident site is declared safe.
2. Immediate resuscitation or transfer of seriously ill or injured patients to hospital should not be delayed by decontamination procedures.
3. Specific treatment is rarely necessary and most patients will require only general supportive measures.
4. Delayed pulmonary oedema may develop up to 24 hours after toxic inhalation. Most patients should therefore be admitted even if they are asymptomatic.
5. Corrosive chemical contamination of the skin and eyes should be treated immediately with copious lavage using water or saline. No attempt should be made to neutralize the caustic agent by using a weak acid or alkali.
6. Appropriate protective clothing must be worn whilst treating and decontaminating casualties from chemical accidents.

General introduction

- **Individual chemical exposure   Major incidents**

Chemicals are an essential and indispensable part of our everyday life. Chemical accidents, however, can occur in nature, industry, and domestic environments. Chemical agents are also used deliberately in warfare. Toxic exposure can occur by skin absorption, inhalation, or ingestion. The emergency medical management of such toxic chemical exposures may include rescue, resuscitation, decontamination, supportive care and specific treatment.

The main hazards posed by the chemical industry are large vapour or flammable gas explosions, fires, and toxic releases. The accidental release of chemicals during distribution by pipeline, water, rail, or road can also have serious consequences.

In the United Kingdom, the most potentially hazardous industrial sites come under the Control of Industrial Major Accident Hazard Regulations 1984 (CIMAH). The implementation of these regulations should include the collaboration of Accident and Emergency departments and the emergency services in planning for local major incidents. Information about local CIMAH sites and chemical industries should be obtained and protocols for emergency responses to chemical accidents should be made available. The necessary facilities in Accident and Emergency departments for decontamination and isolation should be identified. Hospital pharmacies must be made aware of local chemical hazards and the need for antidotes if appropriate. Major incident equipment should include suitable protective clothing and training for on-site work may include the use of breathing apparatus. Frequent rehearsals with the equipment and facilities should be carried out.

Chemical agents may cause multisystem effects. The symptoms and clinical course will vary widely depending on the chemical involved and the concentration, route, and duration of exposure. Incidents may involve mixtures of chemicals which may cause unpredictable toxic effects. Although a few specific antidotes are available the overwhelming majority of casualties from chemical exposure will require symptomatic and supportive care only.

Poisons information centres are essential resources for managing the medical aspects of chemical accidents, can help locate sources of antidote supplies, and can arrange analytical toxicology if necessary (see Appendix).

Chemical agents are discussed in this chapter under skin contamination and toxic inhalation. However, some of these poisonings may occur by both these routes and by other means such as ingestion. Aspects of major incidents and individual chemical exposure are presented.

At the scene

- **Hazard identification Safety protocols Decontamination Resuscitation**

Hazard identification

At the scene, the fire service is responsible for identifying and containing the chemical hazard and rendering it safe. If the incident occurs at the site of storage or processing of a dangerous substance, information can be obtained from on-site safety plans which are required under statutory regulations. Chemicals being transported may be identified from transport emergency (TREM) cards held by the driver or HAZ CHEM codes displayed on the vehicle. The code symbols can be interpreted from a pocket card carried by all members of the emergency services (Fig. 8.1). In addition, the fire service has access to computerized databases such as CHEMDATA which provides information on health hazards, fire fighting techniques, and clean-up procedures for spills. If there is no information as to the nature of the substance, samples should be kept for later identification. CHEMET is a national network which may assist in providing meteorological support in an emergency involving air release of hazardous chemicals.

Safety protocols

Vehicles conveying emergency personnel should approach

Chemical accidents and emergencies • 119

Fig. 8.1 • Examples of hazard warning signs. (Reproduced with permission from Jerrom, D. A. (1990), in *Major chemical disasters: Medical aspects of management* (ed. V. Murray), p. 55. Royal Society of Medicine, London.)

upwind from the hazard and park at a safe distance. Rescue attempts should not be made until the incident site is declared safe by the senior fire officer. Information should be obtained on the degree of protective clothing required and specific treatment protocols which may be necessary. Care must be taken not to step into any spilled liquid, touch substances, or inhale vapours which may be harmful.

Full protective clothing including PVC suits, hoods, boots, gauntlets, visors, masks, and breathing apparatus may be required for entry into heavily contaminated zones. Plastic gowns and gloves are necessary for any chemical decontamination procedure.

Decontamination

The casualty should be moved to a well-ventilated and protected area. Decontamination is usually carried out on-site by the fire service. Running water is usually used and care should be exercised to avoid splashing. A dry decontamination procedure is used for substances that react with water. All clothing exposed to the chemical agent should be removed and chemical solids in contact with the skin wiped away. Clothing stuck to the skin should not be pulled off. Contaminated clothing should either be left at the scene or placed in a plastic bag and sealed. Following decontamination the patient should be wrapped in polythene and blankets to avoid hypothermia. Immediate and thorough irrigation of the eyes is essential if there is evidence of contamination. Care should be taken not to wash the chemical from one eye to the other.

Resuscitation

Immediate resuscitation may be required prior to decontamination and should follow established protocols. Skin or mucosal contact with the casualty must be avoided.

Mouth-to-mouth resuscitation is contraindicated and ventilation with high flow oxygen should be achieved using a bag and mask or endotracheal tube. Intravenous access should be established as required. Urgent transfer of seriously ill or injured patients to hospital should not be delayed by decontamination procedures.

Chemical contamination

- Corrosives   Organophosphate and carbamate insecticides   Mercurials   Chemical burns of the eye

Corrosives

The receiving Accident and Emergency department should be notified and advised if a specific antidote may be required.

Corrosives include the strong acids and strong alkalis and the term is also used loosely to describe a variety of dissimilar agents (Box 8.1).

Box 8.1 Corrosives

Strong acids — Hydrochloric acid
— Hydrofluoric acid
— Nitric acid
— Sulphuric acid

Strong alkalis — Ammonia
— Potassium hydroxide
— Sodium hydroxide
— Calcium hydroxide
— Sodium hypochlorite

Lime or quicklime
Phenol

The common alkalis are widely distributed for household as well as industrial use and are implicated in accidents about 10 times as often as acids. Strong alkalis produce a rapid liquifactive necrosis which deeply penetrates tissues. Strong acids produce a coagulative necrosis which forms a firm protective eschar delaying further caustic damage.

Treatment consists of immediate irrigation with large amounts of water after removal of contaminated clothing.

Dressings should then be applied as for thermal burns. *No attempt should be made to neutralize the caustic acid or alkali by applying a weak alkali or acid.* The heat produced from such an exothermic reaction will do more damage.

Hydrofluoric acid is used extensively in the electronics industry, glass manufacture, and in horticulture. Hydrogen and fluoride ions are released from the acid which in undiluted form can cause severe local tissue damage. The associated pain can be excruciating. Fluoride ions remain active until precipitated by calcium or magnesium released from normal or devitalized tissue. Significant absorption through the skin can cause severe hypocalcaemia and cardiac arrest can occur in extensively burned patients.

Immediate flushing with cold water is recommended followed by the application of ice packs. There is no evidence that calcium gluconate jelly helps in reducing local injury and it provides less pain relief than iced water. If there is evidence of subcutaneous involvement, 10 per cent calcium gluconate (0.5 ml/cm^2 of visible skin involvement) can be infiltrated slowly under regional anaesthesia. This converts the acid into its calcium salt and is not an exothermic reaction.

Lime (calcium oxide) or **quicklime** can produce severe alkali burns because it is readily converted to calcium hydroxide in the presence of water. Skin moisture is sufficient to produce this reaction. It can prove troublesome if particles get into the eye.

Treatment consists of copious lavage. Paradoxically, the use of a small quantity of water may increase the injury because of the exothermic reaction. Irrigation alone may not be sufficient for particulate lime adherent to the surface of the eye and may have to be removed physically after instilling local anaesthetic drops.

Phenol (carbolic acid) may be recognized by its characteristic odour. The skin burns are typically painless because of damage to nerve endings. The skin lesions appear moist, wrinkled, and initially white or yellow in colour. Absorption occurs readily and systemic phenol poisoning can result in hypovolaemic shock, coma, convulsions, haemolysis, and renal failure. The urine is usually grey and may smell strongly of phenol. Despite copious lavage a barrier may form due to coagulative necrosis. Swabbing with polyethylene glycol can reduce local and systemic absorption. If this is not available glycerol is an acceptable substitute.

Organophosphate and carbamate insecticides

Accidental poisoning can occur by percutaneous absorption and inhalation of sprays during use or following ingestion of contaminated foodstuffs. Organophosphates and carbamates inhibit cholinesterases causing accumulation of acetylcholine at all autonomic ganglia, postganglionic parasympathetic nerve endings, neuromuscular junctions, and central cholinergic nerve endings (see Fig. 10.1).

Carbamates have a short-lived affinity for cholinesterase and the complex tends to dissociate spontaneously. The toxicity of organophosphate compounds varies widely, although the duration of cholinesterase inhibition is generally much longer than with carbamates.

Contaminated skin may become red and blistered after a few hours' exposure. The poison may exhibit a garlic odour. Splashes to the eye will cause miosis and disturbed vision on the affected side. Inhalation and ingestion may produce local respiratory and gastrointestinal effects within minutes. Systemic features usually appear in a few hours.

Mild poisoning produces central nervous system effects such as headache, anxiety, tiredness, restlessness, and muscarinic features including nausea, vomiting, abdominal cramps, diarrhoea, sweating, miosis, hypersalivation, and chest tightness. Effects of nicotinic blockade (see Box 10.2) occur with moderate poisoning particularly muscle fasciculation and generalized weakness. Severe poisoning produces widespread flaccid paresis of limbs and respiratory and extraocular muscle paralysis. Ultimately, coma and convulsions develop.

Following exposure, reduction in cholinesterase activity can occur without obvious clinical signs or symptoms. All victims of exposure should therefore be admitted for observation. Blood should be taken for measurement of plasma cholinesterase activity. Activity may be reduced up to 50 per cent in subclinical poisoning and to less than 10 per cent in severe poisoning.

Soiled clothing should be removed and contaminated skin lavaged with water.

A clear airway should be established and high flow oxygen administered. Ventilatory support may be required and should be guided by arterial blood gas analysis. Excessive bronchial secretions must be aspirated and can be facilitated by endotracheal intubation. Fluid losses from vomiting, diarrhoea, and pulmonary oedema should be replaced.

Patients with clinical evidence of poisoning should be treated with intravenous atropine to antagonize the muscarinic effects of acetylcholine. It should be administered in 2 mg doses intravenously every 10–30 minutes until atropinization is achieved (dry mouth, dry flushed skin, and heart rate 70–80 per minute). Pupillary size is an unreliable sign to monitor. Very large doses of atropine, often 30 mg and occasionally much more, may be required in the first 24 hours.

Severely poisoned patients exhibiting muscle fasciculation and twitching or convulsions can be treated with intravenous diazepam.

Oximes (e.g. pralidoxime) can reactivate phosphorylated cholinesterase but are contraindicated in carbamate poisoning. Pralidoxime can be given concurrently with atropine although the dosage of atropine may need to be reduced to prevent atropine toxicity. Pralidoxime is given in a dose of 30 mg/kg by slow intravenous injection. If effective, convulsions and muscle fasciculation should disappear and conscious level and muscle tone should improve. Further doses may be required or a continuous infusion of up to 0.5 g per hour established. Pralidoxime is held in designated centres and can be obtained by contacting poisons information services.

Mercurials

Organic and inorganic mercurials are both used in fungicides. They are readily absorbed through the skin and can be inhaled. Following skin contact they are irritant and corrosive. Inhalation of mercury vapour produces cough, chest pain, and dyspnoea secondary to a pneumonitis. Ingestion of mercuric chloride can cause a severe corrosive haemorrhagic gastroenteritis. Acute systemic poisoning after absorption or inhalation causes headache with a metallic taste in the mouth, hypersalivation, fatigue, impaired memory and mental concentration, dizziness, and ataxia. Severe poisoning can lead to convulsions and renal failure.

Immediate evacuation is required if the atmosphere is contaminated by mercury vapour. Contaminated clothing should be removed and the skin lavaged. Chelating therapy with dimercaprol (British Anti-Lewisite—BAL) is the treatment of choice for inorganic mercurial poisoning when features of systemic poisoning develop. It should be given promptly by deep intramuscular injection in a dose of 2.5–3 mg/kg 4-hourly for 2 days followed by 3 mg/kg twice daily for 8 days. Penicillamine is recommended when organic mercurial poisoning occurs.

Chemical burns of the eye

Regardless of the chemical involved, immediate and thorough irrigation is essential. No time should be wasted in either trying to discover the identity of the chemical or looking for specific irrigants. If no irrigation facility is available at the scene, the entire face should be submerged in a basin of tap water and the eyes opened and closed continuously. An intravenous infusion set attached to a bag of saline is a very convenient and effective means of irrigating the eye both at the scene and in hospital. Irrigation should be continued until the conjunctival sac is neutral on pH testing with universal indicator paper.

Pain and blepharospasm may make irrigation difficult. Warming the irrigation fluid and applying local anaesthetic eye drops may make irrigation more comfortable for the patient. Particulate matter can be removed from the conjunctival fornices by eversion of the lids and swabbing with a cotton wool bud.

After irrigation the visual acuity should be checked in both eyes. On superficial examination the eye may appear deceptively normal and use of a slit lamp gives a more accurate view of any damage. Fluorescein staining is required to reveal the full extent of corneal or conjunctival epithelial loss.

Topical antibiotic (e.g. chloramphenicol) ointments are indicated when corneal epithelial damage is identified. Steroid eye drops have been advocated in the acute stages.

After irrigation all but the most trivial injuries must be referred to an ophthalmologist.

Toxic inhalations

- Respiratory irritants   Carbon monoxide   Cyanide
  Smoke   Hydrogen sulphide   Volatile hydrocarbons

Respiratory irritants

> **Box 8.2 Respiratory irritants**
>
> - Ammonia
> - Chlorine
> - Nitrogen dioxide
> - Hydrogen fluoride
> - Phosgene (carbonyl chloride)
> - Acrolein
> - Sulphur dioxide
> - Hydrogen chloride

Highly soluble compounds (e.g. ammonia, sulphur dioxide, hydrogen fluoride) dissolve in the moisture of the upper respiratory tract. Intense irritation of the nose, eyes, pharynx, and upper airways occurs. Exposure to these agents, therefore, tends to be self-limiting and lung injury occurs infrequently unless very high concentrations are involved or escape is prevented.

Relatively insoluble gases (phosgene, nitrogen dioxide) are not removed in the upper airways and following inhalation produce diffuse alveolar injury. Some agents, such as chlorine, are of intermediate solubility and cause irritation of both the upper and lower respiratory tract. Significant inhalation results in necrosis of the respiratory mucosa and pulmonary oedema. Airway obstruction can result from laryngeal oedema.

Following inhalation of these irritants all but those who

have been minimally exposed should be observed for 24 hours. Patients with pulmonary complications should be given high flow oxygen. Nebulized β_2-agonists may relieve bronchospasm. The pulmonary oedema which is non-cardiac in origin is not responsive to diuretics. In severe cases endotracheal intubation and assisted ventilation may be required. There is no convincing evidence that steroids improve the prognosis.

Carbon monoxide

Carbon monoxide can be produced by incomplete combustion of natural gas (methane), butane, and propane, especially if appliances are poorly installed or maintained, or if there is inadequate ventilation. Other common sources include smoke from all types of fires and car exhaust fumes. The use of paint stripper containing methylene chloride may also lead to carbon monoxide poisoning when it undergoes hepatic metabolism following absorption.

Carbon monoxide has an affinity for haemoglobin 200–250 times that of oxygen. Binding of carbon monoxide to haemoglobin to form carboxyhaemoglobin reduces the total capacity of blood to carry oxygen. However, the tissue anoxia is far greater than would result simply from the loss of oxygen-carrying capacity. This is explained by a left shift and distortion of the oxygen dissocation curve which occurs when one or more carbon monoxide molecules bind to haemoglobin. Carbon monoxide also inhibits cellular respiration by binding to cytochrome oxidase.

The severity of poisoning depends on the concentration of carbon monoxide in the inspired air and the length of exposure. Pre-existing cardiovascular disease, anaemia, or lung disease significantly worsens the prognosis.

Early symptoms of carbon monoxide poisoning include headache, nausea, and dizziness. Confusion and agitation may be seen when cerebral oedema develops in more serious poisoning. Coma eventually supervenes and is associated with increased muscle tone, hyperreflexia, and extensor plantar responses. Papilloedema and retinal haemorrhages may be seen on fundoscopy. Vomiting and diarrhoea have been attributed to bowel ischaemia and when severe may result in haematemesis and malaena. Such a presentation may easily be diagnosed as infectious gastroenteritis rather than carbon monoxide poisoning. Hypotension, tachycardia, arrhythmias, and ischaemic changes on the electrocardiogram can occur.

Arterial blood gas determinations usually show a metabolic acidosis and normal oxygen tension, although the oxygen saturation is reduced. Pulse oximeters are misleading in the presence of carboxyhaemoglobin.

The symptoms, signs, and prognosis of acute poisoning correlate poorly with the carboxyhaemoglobin concentration measured on arrival in the Accident and Emergency department because of the time lapse between exposure and presentation. Nevertheless, carboxyhaemoglobin concentrations of less than 10 per cent are not usually associated with symptoms; greater than 15 per cent suggests significant exposure; and at greater than 60 per cent, coma respiratory arrest and cardiovascular collapse may occur.

Back calculation with a nomogram may be used as a means of estimating concentration at the time of exposure (Fig. 8.2).

The immediate treatment of carbon monoxide poisoning after removal from exposure is to give high concentrations of oxygen. High flow oxygen delivered via a plastic rebreathing mask will deliver a maximum functional inspired oxygen concentration of 50–60 per cent. A leak-tight mask with a non-return valve and oxygen reservoir is recommended for acute carbon monoxide poisoning. Endotracheal intubation and assisted ventilation may be required in comatose patients.

The metabolic acidosis will normally respond to correction of hypoxia but intravenous bicarbonate may be required.

The elimination half-life for carbon monoxide is about 240 minutes when breathing air and 60 minutes when breathing 100 per cent oxygen. Hyperbaric oxygen at 3 atmospheres absolute decreases the elimination half-life to 23 minutes and in addition increases the amount of oxygen dissolved in plasma thereby improving oxygen supply to the tissues. The categories of patients given in Box 8.3 may benefit from hyperbaric therapy but if a suitable chamber is not readily

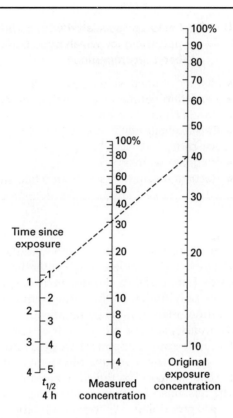

Fig. 8.2 • Nomogram for calculating carboxyhaemoglobin concentration at time of exposure. Time since exposure is given in two scales to allow for the effects of prior oxygen administration on the half-life of carboxyhaemoglobin (the left-hand scale assumes a half-life of 3 hours). For example, if the carboxyhaemoglobin concentration is 30 per cent 1 hour after exposure in a patient receiving oxygen in the ambulance, the concentration at exposure would have been about 39 per cent. [Adapted with permission from Clark, C. J., Campbell, D., and Reid, W. H. (1981). *Lancet*, 1332–5, © Lancet Ltd.]

available administration of high concentrations of oxygen will usually suffice.

Corticosteroids and mannitol have not been shown to be of benefit in cerebral oedema following carbon monoxide poisoning and are not recommended.

> **Box 8.3 Features associated with carbon monoxide poisoning for which hyperbaric oxygen has been recommended**
>
> - History of unconsciousness
> - Persisting neurological deficit or coma
> - Cardiac complications
> - Carboxyhaemoglobin concentrations greater than 40 per cent
> - Pregnant women
> - Recurrent symptoms even at a late stage

Cyanide

Cyanides are used widely in the manufacture of fertilizers, fumigants, rodenticides, and synthetic rubbers. Hydrogen cyanide is released in the fumes from smouldering plastics such as polyurethane. Poisoning from cyanides may follow inhalation, ingestion, or percutaneous absorption.

Cyanides paralyse mitochondrial respiration by binding with the ferrous component in cytochrome oxidase, effectively preventing oxidative phosphorylation. Acute poisoning is characterized by the clinical features of severe hypoxia in the absence of cyanosis. Breathlessness and tachypnoea may be prominent. An odour of bitter almonds may be detected on the patient's breath. Early central nervous system symptoms include anxiety, headache, dizziness, and drowsiness. Coma and convulsions occur in severe cases. Initial tachycardia and hypertension can be followed by hypotension, bradycardia, cardiac conduction defects, and arrhythmias. Arterial blood gas analysis may reveal a severe metabolic acidosis with normal oxygen tension.

At the scene of an incident involving hydrogen cyanide or a liquid cyanide preparation, trained staff wearing protective clothing and breathing apparatus are required to remove the casualty from the contaminated atmosphere.

Intensive supportive treatment is paramount in acute cyanide poisoning. High flow oxygen should be administered. The metabolic acidosis should be corrected by administering intravenous bicarbonate 8.4 per cent titrated against arterial blood gas determinations. Cardiorespiratory support may be required in severe cases.

Contaminated clothing must be removed and any contaminated skin washed with water. When cyanide has been ingested, gastric aspiration and lavage should be performed.

The antidotes for cyanide poisoning are potentially dangerous in the absence of cyanide ions. A patient who has been exposed to hydrogen cyanide gas and who is conscious on admission is unlikely to require antidote therapy. The antidote of choice in the United Kingdom for moderate and severe poisoning (indicated, for example, by depressed consciousness) is dicobalt edetate. Cobalt is a chelating agent and forms a stable inert complex with cyanide. The initial dose is 300–600 mg given intravenously and then a further dose of 300 mg if recovery has not occurred. In the absence of cyanide ions, vomiting, tachycardia, hypertension, chest pain, and facial oedema may occur due to cobalt toxicity.

Smoke

Smoke is a suspension of small particles in hot air and gases. The particulate phase consists of carbon particles coated with combustible products, such as organic acids and aldehydes, which are intensely irritant to the eyes and upper airways and may induce bronchospasm. The gaseous phase consists primarily of carbon dioxide and carbon monoxide although other toxic gases, which vary according to the materials involved and temperature of combustion, may be inhaled (Box 8.4).

Asphyxia may result if the fire uses the available oxygen and carbon dioxide will mechanically displace other gases including oxygen. Carbon monoxide impairs the oxygen-carrying capacity of the blood without causing pulmonary injury (p. 127) and hydrogen cyanide and carbon monoxide paralyse the cytochrome oxidase system. Many of the other toxic gases which may be inhaled cause alveolar injury and will contribute to the asphyxia. Inhaled particles, such as soot, are probably not important in the pathogenesis of pulmonary injury.

132 • Chemical accidents and emergencies

Box 8.4 Materials and products of combustion

Materials	Products of combustion
PVC (polyvinyl chloride)	Hydrogen chloride, phosgene, chlorine
Polyurethane foams	Hydrogen cyanide, isocyanates
Wood, cotton, paper	Acrolein, acetaldehyde, formaldehyde, formic acid
Acrylic	Acrolein
Wool, silk, nylon	Ammonia, hydrogen cyanide
Celluloid, cellulose paints, lacquers	Nitrogen dioxide

Early administration of high concentrations of inspired oxygen is essential. If bronchospasm is present the patient should be given a β_2-agonist such as salbutamol or terbutaline via an oxygen-driven nebulizer.

Clinical evidence of smoke inhalation may also be associated with signs of thermal injury to the upper airway. In patients with one or more of those factors given in Box 8.5, respiratory obstruction may develop rapidly. In such patients early endotracheal intubation by an experienced doctor is essential. In patients in whom complete obstruction has already supervened or intubation is unsuccessful, immediate cricothyrotomy is required followed by formal tracheostomy.

Box 8.5 Clinical features indicating smoke inhalation and thermal injury to the upper airways

- Altered consciousness
- Facial burns
- Singeing of the eyebrows and eyelashes
- Soot in nostrils and sputum
- Hoarseness and stridor
- Dysphagia
- Expiratory rhonchi

Repeated clinical assessment of airway and ventilation is mandatory. This should be supported by frequent measurements of arterial blood gas tensions and peak expiratory flow rates.

The majority of smoke inhalation victims have normal chest X-ray appearances but focal pulmonary infiltrates, diffuse patchy opacities, or changes of pulmonary oedema may be seen. An ischaemic pattern may be revealed on electrocardiogram.

Management of carbon monoxide poisoning, cyanide poisoning, and other toxic inhalations should follow the recommendations previously outlined. Treatment of skin thermal burns are dealt with elsewhere in this series.

Hydrogen sulphide

Hydrogen sulphide is a colourless gas which smells of rotten eggs. It occurs in organic materials and is used extensively in the petrochemical and gas industry. It is irritant to the eyes and respiratory tract. Although detected at first, the distinctive smell may not be noticed after a short time owing to olfactory fatigue. When absorbed in high concentrations there is a cyanide-like action on the cytochrome oxidase system.

The excretion of sulphide can be accelerated by forming sulphmethaemoglobin. Therefore administration of sodium nitrate (10 ml of a 3 per cent solution intravenously) which produces methaemoglobin has been advocated. This should first be discussed with a poisons information centre.

Volatile hydrocarbons

Volatile hydrocarbons are used widely in the manufacturing industry and also in everyday activities (Box 8.6). Poisoning may arise from occupational or domestic accidents or from deliberate inhalation of volatile agents (solvent abuse or 'glue sniffing'). Severe poisoning from accidental inhalation is unusual although the methods used in solvent abuse can enhance the degree of exposure and lead to a fatal dose.

> **Box 8.6 Volatile hydrocarbons**
>
> **Chlorinated hydrocarbons**
> - carbon tetrachloride, trichloroethylene, tetrachloroethylene, trichloroethane — dry cleaning and degreasing agents
> - methylene chloride — paint removers, dyes
>
> **Aromatic hydrocarbons**
> - benzene — paints, varnish removers, detergents, synthetic rubber manufacture
> - toluene — adhesives, acrylic paints
>
> **Paraffin hydrocarbon**
> - *n*-hexane — adhesives
> - butane — lighter refills
>
> **Petroleum and petroleum distillates** — fuels, manufacture of plastics and synthetic rubbers

Volatile hydrocarbons are highly lipid-soluble and affect the lipid component of nerve cell membranes. The clinical features of poisoning are similar to those of alcohol intoxication, with initial central nervous system stimulation followed by depression. Initial effects are of euphoria and exhilaration. Ataxia, dysarthria, impaired judgement, and dizziness may develop with increasing exposure. Eventually, coma and convulsions may supervene.

Inhalation of most hydrocarbons causes only mild respiratory irritation although halogenated hydrocarbons may decompose into hydrochloric acid and phosgene when overheated and produce major pulmonary complications. Severe pulmonary complications will also follow aspiration of petroleum distillates.

Severe hypoxia combined with sensitization of the heart

to endogenous catecholamines by hydrocarbons can cause fatal arrhythmias.

Elimination of hydrocarbons from the body is mainly through the respiratory system, but some solvents are metabolized primarily in the liver. Carbon tetrachloride is particularly hepatotoxic due to a highly reactive intermediate metabolite which produces centrilobular necrosis similar to that seen in paracetamol poisoning. Renal tubular necrosis may also occur. Methylene chloride is metabolized to carbon monoxide (p. 127) but carboxyhaemoglobin levels are rarely high enough to be of clinical concern.

Emergency management following hydrocarbon poisoning is primarily supportive. N-acetylcysteine can be given for potential liver damage in severe poisoning. The degree of central nervous system depression is probably the best indicator of potential hepatic damage.

Injuries due to chemical weapons

- Vesicant agents   Nerve agents   Riot control agents

At least 30 countries world-wide have stockpiles of chemical weapons and there is clear evidence of their recent use against military and civilian targets. Should members of the armed forces fall victim to exposure from chemical weapons, immediate treatment will be provided by the armed forces medical services. Before evacuation to military or civilian hospitals in the United Kingdom all casualties would have been decontaminated and stabilized. Although the principal concerns will be with the late effects of exposure, a knowledge of the agents likely to present the major risk is useful in the initial reception and management of such casualties following evacuation.

Vesicant agents

Mustard gas (sulphur mustard and nitrogen mustard)
Mustard gas is a liquid which gives off a hazardous vesicant vapour. Exposure to the liquid or vapour produces blistering of the skin and damage to the cornea and conjunctivae.

Characteristically, there is a latent period of an hour or more before the onset of symptoms. Crops of blisters may appear at any time during the first 2 weeks following exposure. The blister fluid does not contain sulphur mustard and does not represent a hazard to attendants. Burns may be extensive particularly on the genitalia, perineum, and axillae. Inhalation of the vapour causes damage to the upper respiratory tract with sloughing of epithelium. Absorption of mustards leads to depression of the bone marrow.

Treatment of skin lesions should be the same as for thermal burns. Although the burns tend to be superficial, healing is often slow. The eyes, if involved, require daily irrigation. Mydriatics may ease the pain produced from ciliary spasm but systemic analgesia may also be required. Application of petroleum jelly helps to prevent the lid margins sticking together. Antibiotic drops are useful but local anaesthetic or steroid drops should not be used without expert advice. Dark glasses may be worn for as long as the photophobia is serious. Reassurance is important as eye damage resolves in the vast majority of cases. Damage to the upper respiratory tract should be covered by antibiotics and ventilatory support may be required in severe cases.

Arsenical vesicants

The vesicant effect of arsenical agents (Lewisite) is similar to those of the mustards except that the onset of symptoms is immediate. Systemic arsenical poisoning may follow injury by these agents. Treatment should follow the principles outlined for that of mustard casualties. In addition specific treatment with the chelating agent dimercaprol (BAL) should be instituted. Preparations are available for use locally for the eyes and skin but it should not be used in conjunction with silver sulphadiazine (Flamazine) on burns as BAL ointment chelates the silver. Systemic absorption should be treated with parenteral dimercaprol as for mercurial poisoning (p. 124).

Nerve agents

Nerve agents are organophosphorous compounds which are closely related to the organophosphate pesticides. Nerve agents may be used in conflict in both vapour and liquid phases. Treatment of casualties exposed to these agents is broadly similar to those poisoned with organophosphate pesticides (p. 123).

The efficacy of treatment is greatly enhanced by taking pyridostigmine tablets before exposure and these are available to members of the UK armed forces. Pyridostigmine is a reversible inhibitor of acetylchlolinesterase which, when bound to some of the enzyme, prevents that portion being attacked by the nerve agents. Following exposure to the nerve agent the complex dissociates and a supply of uninhibited enzyme is made available.

Box 8.7 **Chemical weapons**

- Vesicant agents
  — Sulphur and nitrogen mustards
  — Lewisite (arsenical vesicants)
- Nerve agents
  — Organophosphorous compounds (p. 123)
- Cyanide agents (p. 130)
  — Hydrogen cyanide
  — Cyanogen chloride and bromide
- Lung-damaging agents (p. 126)
  — Phosgene
  — Chlorine
- Riot control agents
  — CS (tear gas)

Riot control agents

CS (*ortho*-chlorobenzylidene malanonitrile) tear gas and other irritant but fairly non-toxic 'harassing' agents stimulate afferent nerve fibres in the conjunctiva and respiratory tract and have been widely used to quell civil unrest. Exposure to CS gas causes immediate lacrimation, uncontrollable sneezing and coughing, a burning sensation in the skin and throat, and chest tightness. Vomiting may occur.

Normally, treatment is unnecessary as reversal is rapid on removal from the irritant atmosphere. Severely exposed casualties may develop tracheitis and bronchitis, and local

138 • Chemical accidents and emergencies

irritation of the skin and eyes may require irrigation with water or saline. The attendants should wear gloves.

Further reading

Baxter, P. J. (1991). Major chemical disasters. *British Medical Journal*, **302**, 61–2.

Goulding, R. (1987). Poisoning from chemicals. In *Oxford textbook of medicine*, (ed. D. J. Weatherall, J. G. G. Ledingham, and D. Warrell), (2nd edn), pp. 6.9–13. Oxford University Press.

Langford, R. M. and Armstrong, R. F. (1989). Algorithm for managing smoke inhalation. *British Medical Journal*, **299**, 902–4.

Ministry of Defence (1987). *Medical manual of defence against chemical agents*. HMSO, London.

Meredith, T. and Vale, A. (1988). Carbon monoxide poisoning. *British Medical Journal*, **296**, 77–8.

Meredith, T. J., Vale, J. A., and Proudfoot, A. T. (1987). Poisoning by inhalation agents. In *Oxford textbook of medicine*, (ed. D. J. Weatherall, J. G. G. Ledingham, and D. Warrell), (2nd edn), pp. 6.53–9.

Murray, V. S. G. (ed.) (1989). *Major chemical disasters—Medical aspects of management*. Royal Society of Medicine Services Limited, London.

Murray, V. S. G. and Volans, G. N. (1991). Management of injuries due to chemical weapons. *British Medical Journal*, **302**, 129–30.

Proudfoot, A. T. (1993). Features and management of specific poisons. *Acute poisoning: Diagnosis and management*, (2nd edn), pp. 61–230. Butterworth Heinemann, Oxford.

Chapter 9

Radiation accidents

General introduction	141
Ionizing radiation	141
Types of radiation accident	145
Biological effects of ionizing radiation	147
Acute radiation syndrome	148
The irradiated non-contaminated casualty	151
The contaminated casualty	152
At the scene	152
Arrangements for decontamination in hospital	153
Decontamination of the patient	155
Environmental release: off-site radiation exposure	159
Nuclear weapons accident	162
Further reading	162

Key points in radiation accidents

1 The urgent treatment of serious or life-threatening injuries takes precedence over considerations of irradiation or contamination but it is important to minimize contamination as much as possible.

2 The local radiation protection adviser must be contacted immediately following notification of the impending arrival of a radiation accident victim. He will be responsible for monitoring the level of radioactive contamination and can provide specialist advice.

3 A person who has received a significant dose of external radiation is referred to as an irradiated casualty. Unless the person is also contaminated with radioactive material there is no risk of radioactive contamination of attendants, vehicles, or facilities.

4 A person contaminated with radioactive material is still exposed to radiation and requires urgent and careful decontamination to minimize the radiation dose received. Precautions will be required to minimize the spread of contamination to attendants, vehicles, and treatment facilities.

ns • 141

General introduction

- **Sources of radiation accidents**

Although the use of radiation and radioactive materials has increased in recent years, the probability of having to deal with a radiation accident victim in an Accident and Emergency department is low. Nevertheless, the possibility is very real and there should be a preparedness for such emergencies.

The source of environmental accidents resulting in injury or illness from radiation occurring in this country during peacetime could involve:

- Radiation sources used in industrial, research, educational, or health care establishments.
- Transport of radioactive materials.
- Release of radioactivity from a nuclear reactor within the United Kingdom.
- Nuclear weapons at a Ministry of Defence establishment.
- A major nuclear incident outside the United Kingdom.
- Break up in the atmosphere of a nuclear-powered satellite.

Ionizing radiation

- **Particulate and electromagnetic radiation   Ionizing density   Radiation quantities and units**

Particulate and electromagnetic radiation

Ionizing radiation is radiation that can produce charged particles (ions) in any material it strikes. If such ionization occurs in molecules which are present in living cells, biological damage may result.

Radioactivity is a process whereby atomic changes called *decay* or *disintegration* release energy by emission of ionizing radiation. A radioactive substance is one that emits ionizing radiation and is referred to as a *radionuclide* or *radioisotope*.

Ionizing radiation can be produced from *natural* sources

or by *artificial* means. The natural sources are terrestrial radiation from radionuclides contained in the earth's crust (some of which are incorporated in building materials), cosmic radiation emanating from remote parts of the universe, and internal radiation mainly from potassium 40 which is a radioactive element present in human body fluids. Ionizing radiation that occurs naturally forms *background radiation*.

All types of ionizing radiation can be produced artificially. X-rays are the largest single man-made source of irradiation of the general population. Smoke detectors and gaseous tritium lights are examples of the ubiquitous nature of sources of radioactivity in our lives. Nuclear reactors are not only a source of ionizing radiation but result in the production of large quantities of radioisotopes.

Ionizing radiation may either be *particulate* or in the form of *electromagnetic radiation*. There are five types of ionizing radiation which are important in causing biological damage.

Alpha-particles consist of two protons and two neutrons and are positively charged. They may be regarded as being relatively large. The range of alpha-particles in air is limited to a few centimetres. Alpha-particles cannot penetrate beyond the epidermal layer of skin and therefore pose no external radiation hazard. However, damage can follow ingestion, inhalation, or absorption of alpha-emitting substances through an open wound. Thus alpha-particles constitute an internal hazard only.

Beta-particles are equivalent to electrons and may be positively or negatively charged. Beta-particles can travel several times the distance of alpha-particles in air. With their smaller mass and speed, beta-particles can penetrate skin and subcutaneous tissue if particularly energetic. Clothing will protect covered areas. However, beta-particles can cause primary damage to exposed skin and significant burns can occur. Beta-emitting radioactive substances also pose a hazard if deposited internally.

Gamma-rays are non-particulate and comprise electromagnetic radiation emitted by the spontaneous disintegration of radioactive atoms. Gamma-rays are the most penetrating type of radiation and can travel many metres in air and many centimetres in tissue. Because gamma-rays can travel through the body they are sometimes referred to as *penetrating radiation*. They are the primary cause of the acute radiation syndrome.

X-rays are a more familiar form of electromagnetic radiation. They are produced by bombardment of a positively charged anode with a stream of electrons from a heated filament. Like gamma-rays they can penetrate human tissue. The energy of the X-rays is determined by the voltage to which the electrons are accelerated and the penetrating power of the X-rays increases with increasing voltage.

Neutrons are particulate radiation with a mass approximately one-quarter of an alpha-particle but with no charge. They can travel many metres in air and can have a range in tissue up to many centimetres according to energy. Despite being much larger than beta-particles, neutrons are more penetrating because they are uncharged. Neutrons are produced by only a few elements during radioactive decay, fission, or fusion reactions. In peacetime significant accidents involving neutron exposure would only be found around nuclear reactors or accelerators. Exposure to neutrons may cause damage directly but they can also interact with previously stable atoms and make them radioactive.

Ionizing density

In order to produce ionization, radiation must be energetic enough to dislodge orbital electrons from the atoms of matter through which it passes. The amount of ionization produced will vary with the type of radiation and its energy. Alpha-particles are relatively massive and highly charged and so they readily dislodge orbital electrons. Beta-particles also dislodge electrons directly but are less efficient as they are lighter and carry less charge. Uncharged gamma and X-rays are even less efficient ionizers. They dislodge electrons at high speed and these dislodged electrons in turn pass on to strike other electrons (*secondary ionization*). Neutrons also ionize by secondary means but act by striking the nuclei of atoms. As a result of this, charged particles or gamma-rays

144 • Radiation accidents

are ejected which can then produce further ionization. Radiation can therefore be broadly divided into two groups, *low* or *high linear energy transfer* (LET). Beta-particles, gamma-rays, and X-rays which are sparsely ionizing are classified as low LET whereas neutrons and alpha-particles are classified as high LET.

Radiation quantities and units

The amount of radioactive material present is quantified in terms of *activity*. This is a measure of the number of atoms of the radionuclides which decay to produce the radiation in question in a given time (Table 9.1). The unit of activity is the becquerel which is equal to 1 disintegration per second. However, the activity gives no information on the biological effects which depend on the energy absorbed from the radiation. The unit of *absorbed dose* is 1 gray (Gy) which corresponds to the absorption of 1 joule/kilogram of tissue.

Despite an equivalent dose the biological effects may vary depending on the type of radiation. To compare the effects of different radiations, units of *dose equivalent* are used and is measured in seiverts (Sv). The reference standard for dose equivalent is the effect of sparsely ionizing radiation such as X-rays. One seivert of high or low LET radiation is thus assumed to be equivalent in its effect to one gray of low LET radiation. The use of dose equivalent enables the dosage of all types of radiation to be expressed on a common scale in terms of damaging effects.

The rate at which the dose is delivered is also important in

Table 9.1 • Radiation quantities and units

Quantity	SI unit	Previous unit	Conversion factor
Radioactivity	becquerel (Bq) 1 disintegration sec^{-1}	curie (Ci)	1 Ci = 3.7 × 10^{10} Bq
Absorbed dose	gray (Gy) 1 J kg^{-1}	rad	1 rad = 0.01 Gy
Exposure	coulomb kg^{-1}	roentgen (R)	1 R = 2.58 × 10^{-4} Ckg^{-1}
Dose equivalent	seivert (Sv) 1 J kg^{-1}	rem	1 rem = 0.01 Sv

determining the biological effects produced. Following an accident, acute exposure may lead to the *total radiation dose* being received in a short time compared to chronic exposure when it is protracted over time. The *dose rate* is defined as the dose received per unit of time.

Types of radiation accident

- **External radiation hazard   External and internal contamination   Combined injury**

Exposure to radiation may be from a distant external source or from radiation emanating from internal or external contamination of the body by radioactive materials (Fig. 9.1). An

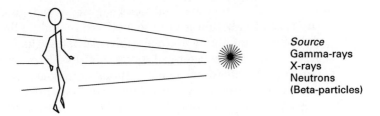

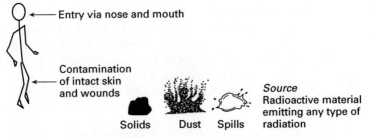

Fig. 9.1 • Types of radiation accident. (Redrawn with permission from Blakely, J. J. A. (1948), *The care of radiation casualties*, pp. 26–7, Butterworth-Heinemann, Oxford.)

external radiation hazard can arise from a source emitting radiation capable of travelling some distance through the air (gamma-rays, X-rays, and neutrons). During exposure, this radiation can be absorbed by the body or it can pass completely through. Beta-particles have to be exceptionally energetic to constitute an external hazard if the source is any distance from the body.

Contamination occurs from radioactive materials in the form of liquids, gases, dusts or solids. The material may become deposited on the skin resulting in external contamination. Internal contamination mainly results from inhalation or ingestion of radioactive substances. Internal contamination can result from exposure on intact skin and wound contamination will provide a further portal of entry. Radioactive contamination of the unbroken skin surface can be transferred on the hands to the nose and mouth and subsequently inhaled or ingested.

After radioactive material has been deposited internally any type of radiation may pose a hazard including the very short-range but highly ionizing alpha-particles. The material may decay or be excreted quite rapidly. However, the radioactive material may be incorporated in bone, liver, and thyroid resulting in significant doses to these organs.

A person exposed to radiation and then removed from the source is referred to as *irradiated* but represents no hazard to attendants. Casualties exposed to neutron irradiation could result in a radiation hazard but would only occur after an accident involving a nuclear reactor or accelerator. *Contaminated* casualties are those with radioactive material on their skin or clothing or who have material deposited internally. The material continues to emit radiation adding to the exposure dose and is also a potential risk to contacts.

Combined injury from external radiation and contamination by radioactive materials can occur and may also be complicated by physical trauma.

Biological effects of ionizing radiation

- **Acute and delayed effects   Whole body and partial body exposure   Somatic and genetic effects   Stochastic and deterministic effects**

The type and severity of harmful effects produced by radiation depends on the dose of radiation. With high doses the effects of exposure, which may lead to death, are noted shortly after exposure. These are known as the *acute effects*. *Delayed effects* may appear following survival from an acute exposure.

A division can also be made between *whole body* (or total body) exposure when all organs are exposed (not necessarily uniformly), and *partial body* exposure when only a few organs are exposed. **An acute dose of ionizing radiation which would have lethal effects if delivered to the whole body may only produce local reactions if delivered to part of the body.** For example, skin burns alone may follow significant doses of radiation to a limb. Internal sources usually give rise to partial body exposure. However, the dose from internal emitters will also be dependent on how the body normally treats the element, e.g. the target organ for radioactive iodine would be the thyroid whereas whole body irradiation would result from tritium exposure.

Radiation-induced biological effects are divided into two main categories. *Somatic* effects are those experienced by the irradiated person. *Genetic* effects are those which become manifest in future generations following germ cell damage. Somatic effects which may develop acutely following total body exposure to relatively large doses of radiation result in the acute radiation syndrome. Delayed somatic effects include the development of cancers, leukaemias, cataracts, and impaired fertility, and follow acute exposure to much lower doses of radiation.

Irradiation during pregnancy may cause somatic damage to the developing fetus. The risk of damage is greatest during the first two months of pregnancy. Irradiation may lead to malformations, mental retardation, or intrauterine death.

Two classes of detrimental effects of radiation are recognized. Cancers, leukaemias, and the genetic effects following irradiation are termed *stochastic*. **There is no threshold for stochastic effects.** They occur with increasing frequency with increasing dose. However, the *severity* does not depend on the dose and these are 'all or none' effects.

The main interest associated with the emergency response to a radiation accident is in the *deterministic* (non-stochastic) effects (Box 9.1). **Deterministic effects are associated with a threshold dose.** They increase in severity with increasing dose and can be induced in any organ given high enough doses.

Box 9.1 Threshold levels for acute radiation exposure below which deterministic effects are unlikely to occur in a normal population

Dose (Sv)	Organ	Effect
0.1	Fetus	Teratogenesis
0.5	Whole body	Nausea
2	Whole body	Early death
2	Lens	Cataract
2	Thyroid	Hypothyroidism
3	Gonads	Sterility
3	Skin	Depilation, erythema
5	Lung	Pneumonitis
10	Lung	Early death

Acute radiation sydrome

- Prodromal phase   CNS effects   Gastrointestinal effects   Haemopoietic effects   Skin effects

When whole body or significant partial body irradiation occurs from exposure to a large acute dose of penetrating radiation (gamma-rays, X-rays, or neutrons) a predictable pattern of effects develops called the acute radiation syndrome. The pattern of illness reflects the sensitivity of the gastrointestinal tract, haemopoietic, and central nervous systems to ionizing radiation.

Prodromal phase

A *prodromal phase* follows shortly after the diffuse cell injury caused by exposure and is probably due to release of toxic breakdown products into the circulation. Symptoms include anorexia, nausea, vomiting, diarrhoea, malaise, fever, and headaches. The higher the dose the earlier the onset and the more severe the symptoms. If vomiting develops within 2 hours the dose has probably been in excess of 2 Sv. Many of these symptoms can be evoked by emotional stress in those at the scene of the accident who have not been exposed. It may be necessary to distinguish this group from those with genuine radiation exposure by appropriate haematological investigation (p. 151).

Below a dose of 4 Sv, symptoms will reach a maximum intensity of about 6 hours after exposure and subside within 2 days. Thereafter there will be a latent period of 2–3 weeks before further symptoms recur. Above 6 Sv, prodromal symptoms may persist and merge into the symptoms of the acute radiation syndrome. Protracted vomiting during the first few days and in particular the onset of diarrhoea indicate a bad prognosis.

CNS effects

Following intense radiation exposure (100 Sv) there is rapid onset of nausea, vomiting, and incapacitation with delirium and ataxia. Within a hour coma and death ensue due to direct radiation effects on central nervous system conduction and cerebral oedema. At lower doses, central nervous system death can occur up to 48 hours after exposure.

Gastrointestinal effects

In the dose range 10–20 Sv death will result around 14 days because of damage to the gastrointestinal tract. The epithelial lining of the small intestine is particularly vulnerable. Reproductive death of stem cells in the crypts results initially in shortening of the intestinal villi. Eventually the lining becomes totally flat drastically reducing the surface area for absorption and also predisposing to bacterial invasion. Injury is heralded by intractable diarrhoea rapidly progressing to watery or bloody stools. Dehydration, circulatory collapse, and overwhelming infection supervene.

Haemopoietic effects

Bone marrow stem cells and lymphocytes are the most radiosensitive cells in the body.

Mature lymphocytes do not need to attempt division before undergoing cell death from radiation damage. The changes in peripheral lymphocyte count over the first 24–48 hours can therefore be used as an early approximate guide to the magnitude of exposure (p. 151). The degree of haemopoietic suppression increases up to about 2 Sv and although maximal at this dose, higher doses will result in earlier falls in cell counts.

The other differentiated peripheral blood cellular elements have normal life spans. However, failure to repopulate peripheral blood from the damaged stem cells by this time results in thrombocytopenia and neutropenia.

Neutrophil counts immediately following exposure are usually high due to a typical stress response, before falling over the next 8 to 9 days. Classically, there is then a slight recovery attributable to an abortive attempt by the bone marrow to repopulate peripheral blood before levels fall to their minimum at about day 30.

At doses of 1–5 Sv the prodromal phase is followed by a period of apparent well-being of several weeks before petechial haemorrhages and mucosal bleeding occur. Overwhelming infection eventually develops usually leading to death within 6 weeks.

Skin effects

Transient erythema will occur within 3 days following doses of up to 4 Sv. At doses higher than this, fixed erythema occurs with hair loss, and above doses of 8 Sv, blistering occurs. Severe and permanent skin reactions occur after such high doses that if the whole body were exposed, the person would die from the acute effects. However, high doses to the skin alone can result from deposition of beta-emitting nuclides on the skin surface.

The irradiated non-contaminated casualty

- **Investigations Prognosis Symptomatic treatment**

The non-contaminated (or decontaminated) radiation accident victim can be handled in the same manner as any other patient presenting to the Accident and Emergency department. In as much as the source of ionizing radiation is not located on or within the body, the victim is not radioactive.

Symptoms and signs of either whole body or partial body exposure, except for very large radiation doses, are unlikely on initial presentation. However, a detailed history and physical examination should still be performed and blood samples taken for laboratory examination (Box 9.2). A full blood count should be obtained immediately after exposure, again 4 hours after the initial estimation and 8-hourly thereafter.

Box 9.2 Initial blood investigations

- Full blood count (differential white cell and platelet count)
- Urea and electrolytes
- Blood group and blood for HLA typing

Within the first 48 hours the prognosis is good if the lymphocyte count remains above 1.2×10^9/litre. Between $0.5–1.2 \times 10^9$/litre the prognosis is guarded and at doses below 0.5×10^9/litre the prognosis is poor.

Radiation-induced nausea and vomiting if present, may be dramatically relieved by ondansetron 8 mg by infusion over 15 minutes and 8 mg by mouth every 8 hours thereafter for up to 5 days.

Appropriate expert medical advice should be sought early.

152 • Radiation accidents

After the acute symptoms have been treated and the initial investigations completed (usually within 2 to 3 days) it may be necessary to electively transfer patients who have received significant doses (greater than 2 Sv) to specialist centres which normally deal with patients who receive therapeutic total body irradiation.

The contaminated casualty

- **At the scene** Arrangements for decontamination in hospital Decontamination of the patient

At the scene

Vehicles conveying emergency personnel to the scene of an accident involving radiation should be positioned upwind of the source. Casualties must be approached from this direction and rescue personnel should wear appropriate protective clothing.

Expert advice may be available at the scene and the presence of monitoring facilities will permit rapid assessment of the extent of contamination. In the absence of such monitoring facilities casualties must be assumed to be contaminated.

If the patient's condition is stable and appropriate facilities and expertise available then decontamination can be performed on-site by the operator's staff.

Serious or life-threatening injury must take priority over decontamination procedures and the patient must be evacuated to hospital without delay. In these circumstances, transfer of contamination can be reduced by placing plastic sheeting above and below the casualty. The sheeting should not be removed until the patient has been admitted to the designated decontamination area in the Accident and Emergency department.

The casualty must not eat, drink, or smoke because these actions might result in internal contamination. These restrictions also apply to emergency services personnel.

A 'contaminated' label should be attached to the casualty

in addition to any other triage labels. If available, dosimetry results obtained at the accident site by the operator should be conveyed to hospital staff.

Radio contact should be made with the receiving Accident and Emergency department to provide advance warning.

Arrangements for decontamination in hospital

Reception and treatment areas

An area should be designated within the Accident and Emergency department to receive and treat contaminated casualties and plans should be shown in the hospital major incident manual. The treatment area should be chosen to allow ease of access and to minimize disruption to the remainder of the department. In considering the choice of area attention should be paid to the ventilation and drainage systems. Ideally, there should be a shower and irrigation facilities. If possible, the designated area should be equipped to allow resuscitation and emergency treatment as well as decontamination to be carried out. An adequate means of communication with the area outside the decontamination zone is required.

Notification

When a call is received advising of the arrival of a contaminated radiation accident victim the information given in Box 9.3 should be obtained.

Box 9.3 Advance notification details for contaminated radiation accident victims

- Name, location, and telephone number of informant
- Time and place of the incident
- Injuries sustained by the casualty
- Type of radiation contamination
- Identity of contaminant
- Expected time of arrival

On receipt of this notification, the Accident and Emergency consultant and the local radiation protection adviser (usually from the medical physics department) should be contacted immediately. A radioactive emergency is likely to attract a lot of media interest. The press officer and an administrator must also be present as soon as possible to attend to journalists and photographers.

Preparation

The designated reception and treatment areas should be cleared of all non-essential portable items. Choose an area which can be readily cleaned afterwards. Covering the floors and walls with polythene or paper is often recommended but is often impracticable. There is usually insufficient warning for such coverings to be laid adequately: in practice, the floor becomes very slippery underfoot when polythene is used and paper tends to tear; and removing coverings without spreading contamination is very difficult. The ventilation system should be switched off and isolated by sealing air-handling grills. The designated treatment area and transit areas should be marked 'radiation area'. Mobile X-ray equipment should be available. A reception trolley or wheelchair should be prepared by covering the top surface with several layers of polythene.

The treatment of a contaminated casualty follows similar principles to that of 'barrier nursing'. Personnel delegated to receive the casualty and to staff the treatment area should put on protective clothing: theatre suits, long plastic aprons, caps, face masks, goggles or visors, rubber boots or overshoes, and two pairs of disposable gloves. The inner gloves should be taped to the sleeves and the outer gloves discarded when contaminated.

Monitoring

A controller should be nominated to record details of all personnel in contact with the contaminated area. The number of personnel involved should be kept to a minimum. Their names should be recorded. A clear control line should be established at the entrance to the decontamination area to differentiate contaminated from uncontaminated zones.

Seriously ill or injured patients should be placed on the prepared trolley and taken to an appropriate resuscitation area which should be, if possible, within the designated area rather than using standard treatment areas. Minor casualties should be transported by wheelchair rather than be allowed to walk to the designated area in order to minimize contamination (Box 9.4).

> Box 9.4 **Principles for managing contaminated casualties**
>
> - Treat life-threatening injuries first
> - Limit radiation dose to victim and personnel
> - Control spread of radioactive contaminants

An area should be designated where the ambulance and crews can be isolated until monitored and decontaminated if necessary.

Following treatment of casualties all instruments and equipment should be monitored for possible contamination. If this is present the item should be placed in polythene bags with radioactive warning labels for specific cleaning. Portable X-ray equipment should be monitored and if necessary decontaminated. All patients and staff must be checked for contamination before leaving the decontamination zone. Finally, all treatment areas, transit areas through the hospital, ambulance and ambulance-hold areas need to be monitored by the radiation physicist before cleared for further use.

Exercising plans

Contingency plans drawn up to deal with casualties contaminated with radioactive substances should be tested regularly. In an exercise, anthracene powder can be substituted to test emergency procedures realistically. Anthracene glows brightly when exposed to an ultraviolet light source, such as a Wood's lamp, but cannot be seen under normal lighting conditions. It also mimics transfer from a 'contaminated' to a 'non-contaminated' area.

Decontamination of the patient

The aim of decontamination is to remove radioactive material and to minimize absorption into the body through the mouth, eyes, nose, and wounds. The distribution and extent of contamination should be monitored prior to removal of clothing and treatment. Where this is not possible in the case of a severely injured patient all clothing should be considered to be contaminated and handled appropriately. Areas of contamination should be mapped out with a skin marker. At the same time a written record should be kept (Fig. 9.2).

Surface decontamination
The patient should be undressed carefully to reduce spread of contaminants. If the patient has been transported between layers of polythene, these should be removed by turning the contaminated *inner* surface *inwards*. Clothes must not be pulled over the head as this may spread contamination to the face and hair. Garments may have to be removed by gently cutting with scissors and turning the contaminated *outer* surface *inwards*.

Fig. 9.2 • Areas of external contamination can be recorded graphically with a note of the type of radiation and radioactivity.

Contaminated clothing belonging to the patient should be placed in polythene bags or containers, sealed, and labelled 'radioactive'.

Radioactive material can usually be removed from the skin surface by washing with soap or cetrimide and warm running water. Exposed areas of skin have priority over areas which are protected by clothing. Gentle scrubbing with a soft brush or surgical sponge may be necessary but it is important to avoid abrading the skin. The nails should be cleaned with a brush for 4 minutes. If monitoring confirms residual contamination the process should be repeated. Decontamination procedures should continue if they are clearly having some effect. If after a number of washes no reduction in level is seen then it is best to stop. In cases where there is significant residual contamination then the area, if practicable, should be isolated in a plastic bag. Specialist advice is required should stubborn contamination remain.

Box 9.5 Rules of priority for external decontamination

- Remove clothing
- Cover uncontaminated wounds
- Decontaminate in the following order:
  — contaminated wounds
  — nose, eyes, mouth
  — near nose, eyes, mouth
  — near uncontaminated wounds
  — 'hot spots' on skin
  — other areas of skin
- Finally, contaminated wounds can be debrided if necessary

The face should be carefully swabbed using saline with the patient's eyes closed and ears plugged. Swabs should be kept for measurement of radioactivity. The mouth should be cleaned using a mouthwash and a soft toothbrush with care to avoid swallowing any fluid. The ears should be cleaned with a pledget or irrigated with saline, assuming the tympanic membranes are intact. The eyes should be irrigated from the medial sides outwards to avoid drainage into the nasolacrimal duct. Nasal swabs should be taken and kept separately for subsequent analysis. The patient should then blow his nose into tissues (which should be retained for monitoring). If contamination remains the nose should be irrigated with a small amount of normal saline.

The hair should be washed with shampoo. If contamination persists clipping may be required but the head should not be shaved.

All samples must be placed in *separate*, labelled ('radioactive') containers/double bags that specify name, date, time, and site of sampling.

When monitoring confirms that all contamination has been removed the patient should then be treated as for an irradiated but non-contaminated casualty, unless internal contamination has already taken place. (See Box 9.5 for sequence of priorities for external decontamination.)

Wound decontamination

Wounds should be decontaminated before intact skin. Each wound should be draped with a waterproof material to limit the spread of radioactivity. Any obvious large pieces of contaminant material should be removed with forceps. The wound should be irrigated with copious amounts of saline. The irrigation fluid should be collected and checked with a radiation monitor. The wound should also be monitored after each irrigation but the contaminated drapes must be removed for accurate results. Decontamination may be assisted by encouraging the wound to bleed. After the wound has been decontaminated it should be covered with a waterproof dressing while the remaining skin is decontaminated. If these measures are unsuccessful and the contamination level is still significantly high, debridement may be necessary. This should be guided by expert advice. Debrided or excised tissue should be kept for assessment. Without sophisticated specialist wound probes it is impossible to be sure that decontamination of deep wounds has been achieved. All such patients therefore require admission for repeat wound probing and also for monitoring of uptake and internal contamination.

Internal contamination

Initial monitoring may reveal contamination of the mouth, nose, or pharynx. In these circumstances it must be assumed that radioactive material has been ingested or inhaled and that internal contamination has occurred. If internal contamination is suspected, samples of urine, faeces, and vomitus should be kept. Subsequent incorporation can occur within minutes or hours with some radionuclides, so urgent measures to prevent incorporation may be necessary. Methods to increase elimination or excretion, prevent absorption, and use of blocking or chelating agents should be guided by expert medical opinion. The hazards of stable iodine are so miniscule and the window of opportunity for its successful use so small that, in any case where contamination with radioactive iodine is known to have taken place, stable iodine should be given for prophylaxis—by giving potassium iodate tablets or Lugol's iodine solution on the tongue. Known iodine allergy is a contraindication to stable iodine administration.

Specialist referral will be necessary to determine the extent of internal contamination.

Environmental release: off-site radiation exposure

- **Off-site radiation exposure** **Urgent countermeasures**
  Emergency reference levels

Off-site radiation exposure

Irradiation hazards can develop 'off-site' and involve individuals not in the immediate vicinity of the accident site. More widespread environmental radiation hazards could arise from:

1 Exposure to gamma-radiation from a radioactive cloud following substantial airborne release or from gamma-emitting radionuclides deposited on the ground or in dwellings.

160 • Radiation accidents

2 Exposure to external gamma- or beta-radiation from radionuclides contaminating the body, clothing, or possessions.
3 Internal exposure to alpha-, beta- or gamma-radiation following inhalation or ingestion of radioactive substances as a result of direct atmospheric or environmental contamination or, subsequently by radioactive material in water or food.

Urgent countermeasures

Following an accidental release of radioactive material into the environment emergency plans include measures to protect the public from the radiation hazard. Such urgent countermeasures include:

Sheltering Members of the public in the relevant area may be advised to stay indoors to reduce radiation exposure.

Evacuation Evacuation of a limited area downwind of the accident site may be carried out. Emergency medical and health services would be responsible for the arrangements to evacuate patients from hospitals and other residents in the community who cannot be accommodated or transported by local authority services.

Administration of stable iodine Where material escaping after an accident contains radioactive iodine, inhalation and subsequent incorporation in the thyroid will result in a high dose localized in the gland. Uptake can be blocked by the administration of stable iodine. This should be administered as early as possible but useful blocking can still be achieved several hours after exposure (60–70 per cent at 3 hours, 50 per cent at 5 hours). Arrangements exist at major nuclear establishments for stable iodine tablets to be issued to the public. The tablets are usually distributed by the police.

The principles underlying countermeasures are that the benefits to those concerned should outweigh the risks. The decision to invoke countermeasures are governed by emergency reference levels (ERLs). The projected radiation dose of the population at risk is assessed from measurements of radioactivity in the field or from measurements from samples of material obtained from the environment and action then determined according to upper and lower ERLs (Box 9.6).

Emergency reference levels

> **Box 9.6 Examples of emergency reference levels for invoking countermeasures**
>
> | | Dose equivalent levels (mSv) | |
> |---|---|---|
> | | *Lower* | *Upper* |
> | **Sheltering** | | |
> | — Whole body | 3 | 30 |
> | — Thyroid/Lung/Skin | 30 | 300 |
> | **Evacuation** | | |
> | — Whole body | 30 | 300 |
> | — Thyroid/Lung/Skin | 300 | 3000 |
> | **Stable iodione** | | |
> | — Thyroid | 30 | 300 |

Action is not recommended when the radiation dose to the individual, which could be averted if the countermeasure were to be invoked, is projected to be less than the lower ERL. Between this and the higher ERL, action depends on circumstances and beyond the upper ERL the introduction of countermeasures is virtually certain. Implementation of one or more of these countermeasures depends not only on the nature of the accident and its time phase but also on specific local conditions such as population size and weather conditions.

On-site and off-site monitoring at a fixed major radiation accident site would normally be carried out by the operator's staff. Off-site monitoring, following, for example, a transport accident, would be performed by health service teams. In the early phase following an accident where ERL-based emergency countermeasures are being taken, requests for personnel monitoring should be deflected until the resources become available. However, evacuated patients, members of the public, and emergency services personnel engaged in the evacuation will require monitoring and, if necessary, decontamination.

Nuclear weapons accident

- **Non-nuclear high explosives   Plutonium**

A nuclear (fission) weapon basically consists of uranium and plutonium, a conventional high explosive, and the necessary triggering device to detonate the weapon. A nuclear detonation can be produced only upon proper functioning of the weapon in the normal sequence of arming and firing. The greatest accidental threat is not from nuclear detonation but from the non-nuclear high explosives and the plutonium. The explosives may ignite, burn, and, in some cases, explode in the event of a fire. When associated with a fire, metallic plutonium may burn producing radioactive plutonium oxide particles. In addition, plutonium would be spread widely following detonation of the conventional explosives. Plutonium presents serious hazards if inhaled, ingested, or deposited in wounds. Plutonium is not a radiation hazard if it remains outside the body because it is an alpha-emitter. Patients contaminated in nuclear weapons accidents can be handled in a similar manner to other contaminated cases. Alpha-monitoring will be required and faecal analysis would be of special importance.

Consideration should be given to the use of a chelating agent following inhalation or ingestion of plutonium. Public protection should be based on the considerations already described in relation to emergency reference levels.

Further reading

Department of Health (1989). *Health services arrangements for dealing with accidents involving radioactivity.* HC(89)8. HN(FP)(89)8.

Heaton, B., Matheson, A. B., and Page, J. G. (1990). Radioactive

substances decontamination exercise. *British Medical Journal*, **300**, 1121–2.
Scottish Home and Health Department (1990). *Planning guidance for the NHS in Scotland—Incidents involving ionizing radiation*. SHHD/DGM (1990) 30.
Smith, J. and Smith, T. (1981). Medicine and the bomb. *British Medical Journal*, **282**, 771–4, 844–6, 907–8, 963–5.
Stockholm International Research Institute (1981). *Nuclear radiation in warfare*. Taylor and Francis, London.
Stott, A. N. B. (1979). The management of radiation accidents. *Practitioner*, **223**, 54–8.
World Health Organization (1987). *Nuclear power: Accidental releases–practical guidance for public health action*. WHO regional publications, European series No. 21.

Chapter 10

Poisoning by plants and fungi

General introduction	167
General aspects of management	167
Toxic plant ingestions	171
Poisonous fungi	179
Plant dermatitis and phototoxicity	185
Further reading	186

Key points in poisoning by plants and fungi

1. Serious poisoning from ingestion of poisonous plants or fungi is rare in the United Kingdom, with only two deaths in a 15-year period.
2. The initial approach to the management of toxic plant and mushroom ingestion is the same as for any other poisoning.
3. Physical examination may reveal certain characteristic toxic syndromes. This may help categorize plant ingestions which can assist in management and anticipation of complications before definitive identification can be made.
4. Following mushroom ingestion the interval between ingestion and onset of symptoms is crucial in distinguishing between non-serious and potentially fatal episodes.
5. Symptomatic and supportive care is the mainstay of treatment. The few antidotes which are available should be used with caution.
6. Do not expect poisons information services to identify plants and mushrooms from descriptions given over the telephone. These must be identified prior to the enquiry.

General introduction

Poisoning from plants and mushrooms in the United Kingdom is not a serious problem in terms of mortality with only two deaths occurring in a period of 15 years. However, approximately 6000 inquiries related to ingestion of plant material are received by poisons information centres every year.

The definition of a poisonous plant is not as simple as it may first seem. A plant may be considered poisonous to man if adverse effects are produced. However, not all such plants are poisonous in all their parts or at all stages of their growth. In addition, the ingested amount which is required to produce harmful effects varies widely between plants.

Toxic plants are ingested mostly by curious children attracted to bright colourful berries or seeds. Mistakes may be made by ramblers who pick and eat wild varieties of mushrooms and plants and by amateur herbalists looking for natural remedies. Careless selection of the type and quantity of plants and mushrooms for use as 'recreational' drugs can also lead to poisoning.

General aspects of management

- **Clinical examination   Identification   Supportive treatment   Emptying the stomach   Preventing absorption   Antidotes   Poisons information centres**

Clinical examination

A detailed history following toxic plant and mushroom ingestion is crucial in distinguishing a potentially lethal from a mild poisoning episode. Important aspects of the history include:

- Time of ingestion
- Interval between ingestion and onset of symptoms
- The amount and the parts consumed
- The number and different varieties consumed

- Alcohol intake prior to onset of symptoms (mushroom poisoning)
- Method of preparation, e.g. cooking
- The symptoms, if any, of others eating the same preparation

The physical examination may reveal certain characteristic toxic syndromes (e.g. anticholinergic effects induced by some plant alkaloids) which may help categorize the ingestion into one of several broad groups. This may assist both in the management and anticipation of complications. Therefore, it may be possible to institute appropriate care before definitive identification of the ingested plant or mushroom can be obtained.

Identification

Amateur attempts to distinguish species following a poisoning episode may be dangerous and when required should be left to a trained botanist and mycologist. It is essential to obtain as much of the original plant or mushroom as possible for identification. Care should be taken to avoid damaging the specimen which should be placed in a paper bag (not plastic) and refrigerated. Specimens should be clearly labelled so no one else eats it! A sample of emesis if available should be kept refrigerated—following mushroom poisoning, spores or hyphae may be the only remnants of the original ingestion.

Supportive treatment

The initial approach to management of toxic plant or mushroom ingestion is the same as for any other poisoning. Resuscitation should commence immediately in unresponsive patients, including appropriate attention to airway care and ventilatory support.

Intravenous access should be established in severe poisoning as fluid losses from vomiting and diarrhoea can be considerable. Meticulous control of fluid and electrolyte balance is a crucial part of supportive care especially in children.

Convulsions may complicate poisoning with a wide variety

of plants and mushrooms. Most are short lived. However, intravenous diazepam may be titrated against the response in protracted or recurrent fits.

Electrocardiographic monitoring is required in toxic ingestions with potential for cardiac arrhythmias. The arrhythmias produced in severe poisoning episodes by the toxic principles of plants and mushrooms are complex. As a general rule, such arrhythmias do not require treatment with antiarrhythmic drugs unless associated with hypotension. Insertion of a pacing wire may be more helpful in severe cases. Attention to supportive measures ensuring adequate oxygenation, correction of acid–base disturbances, hypokalaemia, and hyperkalaemia should improve the situation in the majority of cases.

Baseline urea and electrolytes, full blood count, and blood gas determinations may be helpful in serious or potentially serious poisonings.

Emptying the stomach

The stomach should be emptied if the patient presents within 4 hours of ingestion, unless identification of the mushroom or plant indicates that there is no risk of serious poisoning.

If vomiting has not been induced by the poison itself emesis should be induced by syrup of ipecacuanha (30 ml for adults and 15 ml for children) followed by 200 ml of water. If ineffective, this dose can be repeated at 20 minutes. In unresponsive patients or those with a depressed gag reflex, a cuffed endotracheal tube should be placed before gastric lavage is performed with a large-bore stomach tube.

Preventing absorption

Activated charcoal is particularly effective in absorbing toxic plant alkaloids. Conscious patients should be able to drink an oral loading dose of 50 grams. If this is impossible or vomiting is a problem activated charcoal should be given down a nasogastric tube. If administration is delayed for an hour or more after ingestion the ability of activated charcoal to prevent absorption is greatly reduced. However, activated charcoal may also be given regularly over a period of a few days to interrupt the enterohepatic circulation of some plant (such as digitalis) and mushroom toxins which have long plasma half-lives.

Haemodialysis and charcoal haemoperfusion have not been shown to be effective in removing any of the known plant or mushroom poisons.

Antidotes

There are few specific antidotes for toxic plant and mushroom ingestions and the toxins for which they could be used often produce complex and rapidly changing clinical pictures. Therefore, early administration of such 'antidotes' may exacerbate rather than improve the clinical situation. Symptomatic and supportive care is the mainstay of treatment.

Poisons information centres

If there is any doubt about the nature or toxicity of the plant or mushroom poisoning, information and advice can be obtained by contacting a poisons information centre. The telephone numbers of UK centres are given in the Appendix.

Do not expect poisons information services to identify plants and mushrooms from descriptions given over the telephone. These must be identified prior to the enquiry. Books which may be useful for identification are included in the list of further reading at the end of this chapter. A computer software package ('Plato') for identification of plant material is also being developed by the London Centre of the National Poisons Service, and botanists at Kew Gardens. (Box 10.1 indicates plants that are non-toxic.)

Box 10.1 **Plants that are non-toxic, but may cause mild gastrointestinal symptoms following ingestion**

Barberis spp. (Barberry)
Begonia spp.
Chlorophytum spp.
  (Spider plant)

Lunaria annua (Honesty)
Mahonia spp.
Muscari spp.
  (Grape hyacinth)

Cotoneaster spp.
(see p. 176)
Crataegus monogyna
(Hawthorn)
Cyperus alternifolius
(Umbrella plant)
Ficus spp. (Rubber plant/
Weeping fig)*
Fuchsia spp.
Geranium spp.*
Hypericum spp. (St.
John's wort)*
Impatiens spp. (Busy
Lizzie)
Ilex spp. (Holly)
Lonicera spp.
(Honeysuckle)
Pyracantha spp.
(Firethorn)
Rosa spp. (Rose/
Rosehips)*
Saintpaulia ionantha
(African violet)
Sambucas spp.
(Elderberry)
Skimmia japonica
Solanum pseudocapsicum
(Winter cherry)
Sorbus spp. (Rowan/
Mountain ash)
Tradescantia spp.
(Wandering jew)
Yucca spp.

*These plants may cause a mild dermatitis.

Toxic plant ingestions

- **Nicotinic effects   Antimuscarinic effects
Cardiovascular disturbance   Cyanide poisoning
CNS disturbance   Gastrointestinal irritation**

In the United Kingdom most poisonous principles associated with plant poisoning are *alkaloids* and *glycosides*. Many such toxic principles produce their effects by mimicking or blocking the actions of neurotransmitters (Fig. 10.1).

The following common and important examples of toxic plant ingestions in the United Kingdom have been categorized on the basis of the clinical effects they commonly produce and the toxic principles involved. However, many plant toxins have complex actions on multiple organ systems and the clinical presentation may vary depending on the quantity ingested. Specific treatments where appropriate are given but emphasis is again placed on the requirement for supportive care previously outlined.

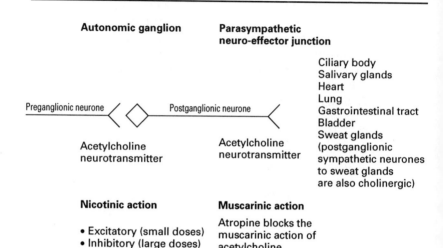

Fig. 10.1 • Autonomic neurotransmission. *Note*: the neurotransmitter at neuromuscular junctions in skeletal muscle is also acetylcholine and acts on nicotinic receptors. Cholinergic receptors are also found in parts of the CNS.

Plants with nicotinic effects

Pyridine–Piperidine alkaloids (e.g. nicotine, coniine)
- Hemlock (*Conium maculatum*)—All parts, particularly seeds and young leaves

The alkaloids in this group characteristically cause initial nicotinic stimulation of autonomic ganglia, neuromuscular junctions, and central nervous system cholinergic receptors, followed by depression of transmission as blockage supervenes.

Poisoning is heralded by nausea, vomiting, and abdominal cramps. Transient signs of parasympathetic and central nervous system stimulation are then followed by depressed conscious level and respiratory paralysis (Box 10.2).

Treatment of the initial bradycardia with atropine should be resisted as there is a natural progression to tachycardia. Temporary pacing is preferable if the patient is severely compromised.

> **Box 10.2 Nicotinic effects**
>
> **Initial transient nicotinic stimulation**
> — Miosis
> — Hypersalivation
> — Bradycardia
> — Sweating
> — Increased respiratory rate
> — Agitation
>
> **Subsequent nicotinic blockade**
> — Mydriasis
> — Tachycardia
> — Hypotension
> — Limb weakness
> — Respiratory paralysis
> — Coma
> — Convulsions

Quinolizidine alkaloids (e.g. cystosine, lupinine)
- Laburnum (*Laburnum anagyroides*)—All parts especially, seeds, pods, and leaves when ripe
- Broom (*Cystisus scoparius*)—Seeds
- Lupins (*Lupinus* spp.)—All parts, especially seeds

The seeds of the laburnum are often eaten by small children and are traditionally regarded as poisonous but few children develop any symptoms and serious poisoning is very rare. The one death reported in the United Kingdom in the last 50 years was a psychiatric patient who ate a very large quantity of seeds.

Treatment of laburnum ingestion is usually unnecessary. Nausea, vomiting, drowsiness, and abdominal pain occasionally develop and require observation and supportive treatment. Poisoning from broom or lupin only occurs following ingestion of large amounts.

The quinolizidines resemble nicotine in their actions (Box 10.2).

Plants that cause antimuscarinic effects

Tropane alkaloids (e.g. atropine, hyoscine, hyoscyamine)
- Deadly nightshade (*Atropa belladonna*)
   — All parts, especially berries
- Thornapple or Jimson weed (*Datura stramonium*)
   — All parts, particularly seeds

The tropane alkaloids are often called belladonna alkaloids. The antimuscarinic effects of tropane alkaloids (Box 10.3) result from competitive inhibition at neuro-effector junctions. The effects of the tropane alkaloids on the central nervous system are less well understood.

Box 10.3 Antimuscarinic syndrome

- Dry mouth
- Warm dry skin
- Mydriasis
- Tachycardia
- Urinary retention
- Agitation
- Confusion
- Visual hallucinations
- Delirium
- Drowsiness
- Coma

Treatment is supportive, as for tricyclic antidepressant poisoning, with ventilation if necessary. Diazepam orally is helpful for sedating very agitated patients. There is no evidence that treatment with the cholinesterase inhibitor physostigmine is superior to supportive treatment and its use is associated with dangerous side-effects.

Plants that cause cardiovascular disturbance

Cardiac glycosides
- Foxglove (*Digitalis purpurea*)—Leaves and flowers
- Lily of the valley (*Convallaria majalis*)—Leaves and flowers

Poisoning by plants and fungi • 175

- Oleander (*Nerium oleander*)—All parts, especially seeds

Digitalis is the most renowned member of this group although the other members are more commonly implicated in plant poisoning.

Toxicity is due to inhibition of the cellular sodium–potassium transport mechanisms. Symptoms typically begin 1–2 hours after ingestion (Box 10.4).

Box 10.4 Features of cardiac glycoside poisoning

- Nausea, vomiting, and abdominal colic
- Diarrhoea
- Visual disturbance
- Cardiac arrhythmias
- Dizziness, disorientation, and stupor

Arrhythmias are potentially the most dangerous sequelae and the electrocardiogram may show sinus bradycardia, atrioventricular conduction defects, extrasystoles, and tachyarrhythmias. The central nervous system disturbance tends to be minimal until late during the clinical course. The bradycardia usually responds to treatment with atropine but occasionally cardiac pacing is required.

Hyperkalaemia is common in severe poisoning and a metabolic acidosis may also be present. The hyperkalaemia should be corrected by giving 10 units of soluble insulin and 50 ml of 50 per cent dextrose. Intravenous sodium bicarbonate (1 mmol/kg) may be given initially for a metabolic acidosis. Repeated oral doses of activated charcoal can be administered to absorb any toxin remaining in the gut and to interrupt the enterohepatic circulation. Digoxin-specific Fab antibody fragments can be used in cases of severe intoxication.

Viscotoxins (e.g. tyramine, phenylethylamine)
- Mistletoe (*Viscum album*)—Berries

Viscotoxins are long chain polypeptides some of which are structurally and pharmacologically similar to the cardiotoxin of cobra venom. They produce depolarization of muscle cell membranes by binding to sites normally occupied by calcium.

Children normally only eat a few berries and symptoms are rare. Vomiting, abdominal cramps, and diarrhoea are the most common features, and muscle weakness, bradycardia, and circulatory failure very rarely develop.

The stomach should be emptied if more than 10 berries have been ingested and observation is required. Calcium gluconate might be helpful if severe circulatory collapse occurs and atropine may be required for bradycardia associated with hypotension.

Plants that cause cyanide poisoning

Cyanogenic glycosides (e.g. amygdalin)
- Cotoneaster (*Cotoneaster* spp.)—Berries
- Elderberry (*Sambucus* spp.)—Unripe berries
- Apricot, peach, plum, and cherry (*Prunus* spp.)—Kernels

Following ingestion, hydrocyanic acid may be liberated from the glycoside if the seeds are disrupted (Box 10.5).

Ingestion of cotoneaster berries is one of the most common of plant enquiries to the National Poisons Information Service. *Most children will have no symptoms. Others may vomit or develop abdominal pain or diarrhoea. Only rarely are sufficient quantities ingested to cause severe symptoms.*

Box 10.5 Clinical features of cyanide poisoning

- Agitation
- Tachycardia
- Dyspnoea
- Increased respiratory rate
- Chest tightness
- Hypotension
- Impaired consciousness
- Coma

The stomach should be emptied if a large number of berries have been ingested. In cases of severe poisoning, a clear airway should be established and high flow oxygen provided. Ventilatory support may be required. If gastric lavage is performed it should be preferably carried out using 5 per cent sodium thiosulphate and 200 ml of 25 per cent sodium thiosulphate may be left in the stomach following lavage. Lavage should be performed with water in the normal way if sodium thiosulphate is not immediately available.

Antidotes are potentially dangerous in the absence of cyanide ions. Therefore the diagnosis should be beyond doubt and of a severe nature (coma, respiratory depression, profound hypotension) before an antidote is given. The antidote of choice is dicobalt edetate 600 mg intravenously (see Chapter 8, p. 131).

Plants that cause central nervous system disturbance

Resins (e.g. cicutoxin, oenanthotoxin)

- Cowbane (water hemlock) (*Cicuta virosa*)—All parts, especially roots
- Hemlock water dropwort (*Oenanthe crocata*) — All parts, especially roots

Resins are a group of highly toxic unsaturated aliphatic alcohols. The CNS effects may be due to over-stimulation of central cholinergic pathways.

Hemlock water dropwort is probably the most poisonous plant in the United Kingdom. The roots may be mistaken for edible parsnips. A single rhizome may cause fatal poisoning.

Following ingestion, the onset of symptoms is dramatic and occurs within 1 hour. Initial symptoms include malaise, nausea, vomiting, hypersalivation, abdominal cramps, and palpitations, followed by the characteristic features of CNS involvement (Box 10.6).

Box 10.6 **CNS effects of hemlock water dropwort poisoning**

- Extensor muscle spasms
- Trismus
- Opisthotonos
- Convulsions

A rise in aspartate aminotransferase levels and hypoprothrombinaemia, reflecting hepatocellular damage rarely develop.

There is no specific treatment for hemlock water dropwort poisoning. The stomach should be emptied and convulsions controlled.

Plants that cause gastrointestinal irritation

Glycoalkaloids (e.g. taxine, aconitine, solanine)
- Yew (*Taxus baccata*)—All parts, especially seeds
- Monkshood (*Aconitum napellus*)—All parts, especially roots
- Woody nightshade (*Solanum dulcamara*)—All parts, especially unripe berries
- Black nightshade (*Solanum nigrum*)—All parts, especially berries
- Potatoes (*Solanum tuberosum*)—'Greened' or sprouted tubers

The intact alkaloid is a severe irritant to mucous membranes and gastrointestinal tract (Box 10.7).

Box 10.7 **Glycoalkaloid poisoning**

- Burning of mouth and throat
- Swelling of the lips and tongue
- Vomiting
- Abdominal cramps
- Diarrhoea

Upon hydrolysis, the glycoalkaloid yields a sugar and an alkamine. Following absorption, the alkamine accounts for effects primarily on the cardiovascular and central nervous systems. Bradycardia and hypotension develop and in severe cases delirium, coma, and convulsions occur.

Atropine intravenously may be helpful for the bradycardia, and hypotension will usually respond to fluid administration. There are at least two reports of yew berry ingestion causing cardiac arrhythmias, one of them suggesting that digoxin-specific Fab antibody fragments might work.

Coumarine glycosides (e.g. mezerein, daphnetoxin)
- Horse chestnut (*Aesculus hippocastanum*)—All parts
- Spurge olive (*Daphne mezereum*)—Berries
- Spurge laurel (*Daphne laureola*)—Berries

Coumarines act similarly to the irritant glycoalkaloids and treatment is supportive.

Oxalates (e.g. calcium oxalate)
- Araceae (Dieffenbachia, Monstera, Philodendron)
—All parts
- Black bryony (*Tamus communis*)—Berries

The Araceae are common ornamental household plants such as the Swiss cheese plant (Monstera). All parts of the plant contain needle-like crystals of calcium oxalate. When any part of the plant is chewed, the sharp crystals disrupt the mucous membrane permitting entry of a protein with histamine-like properties. Severe burning of the lips and mouth and hypersalivation occur. Acute pharyngeal oedema can lead to airway obstruction.

In addition to purgative effects, the scarlet berries of black bryony contain an unidentified irritant which causes blistering on skin contact.

Symptomatic relief of the mucosal irritation may be obtained from antacids or topical soothing preparations.

Poisonous fungi

- Speed of onset of symptoms   Toxic principles and effects

The number of mushrooms which cause serious poisoning is small and 90 per cent of fatalities are due to one species, *Amanita phalloides*.

The interval between ingestion and onset of symptoms is crucial in distinguishing between non-serious and potentially fatal poisoning. Patients who have ingested mushrooms usually have a self-limiting gastrointestinal upset within 2 hours after eating the fungus. Delayed onset of symptoms (6–8 hours) usually indicates more serious and potentially life-threatening poisoning, although a mixture of species ingested may confuse the clinical picture.

The following examples commonly implicated in mushroom poisoning are grouped according to the speed of onset of symptoms and the toxic principles involved. Specific treatments are given for severe cases. However, the general principles of symptomatic and supportive care outlined on pp. 167–70 should be followed.

Gastrointestinal irritants (numerous fungal species)

Interval: Less than 2 hours.

The most common adverse reaction to eating mushrooms is gastrointestinal irritation with nausea, vomiting, abdominal pain, and diarrhoea. The toxins involved remain largely undetermined. Raw mushrooms produce symptoms more frequently than cooked mushrooms.

The illness is usually self-limiting with symptoms resolving in 3–4 hours. Treatment is supportive with correction of fluid and electrolyte imbalance if gastrointestinal losses are severe. Children are particularly vulnerable.

Inocybe and *Clitocybe* species

Interval: 15–30 minutes.

The toxin is thermostable and not destroyed by cooking. These species contain large quantities of muscarine which produces a typical clinical syndrome (Box 10.8).

Effects on the central nervous system are minimal as the toxin does not cross the blood–brain barrier.

Induced emesis and gastric lavage must be performed with care as these procedures may further exacerbate the vagal effects of muscarine. Atropine may be considered if there is bradycardia and hypotension. The initial dose is 0.6 mg intravenously in adults and 0.02 mg/kg in children. Further doses are titrated against the response until normal sinus rhythm returns. Miosis may last for many hours after adequate atropinization.

> **Box 10.8 Muscarinic effects following mushroom poisoning**
> - Sweating
> - Hypersalivation
> - Vomiting
> - Miosis
> - Abdominal cramps
> - Bronchospasm
> - Diarrhoea
> - Bradycardia
> - Hypotension

Psilocybe and *Panaeolus* species

Interval: less than 2 hours.

The 5-hydroxytryptamine derivatives (5-HT), psilocybin, and psilocin are the toxins implicated.

These mushrooms have been used as 'recreational' drugs. So-called 'magic mushrooms' include the liberty cap (*Psilocybe semilanceata*). Psychomimetic symptoms similar to intoxication with lysergic acid diethylamide (LSD) predominate (Box 10.9).

> **Box 10.9 Psychomimetic features of *Psilocybe* spp. poisoning**
> - Hallucinations
> - Mydriasis
> - Sinus tachycardia
> - Diastolic hypertension
> - Flushing of upper trunk and face
> - Hyperreflexia

Transient psychotic and more prolonged schizophrenic-like states have been reported.

Emptying the stomach is probably unnecessary. Sedation may be required with titrated (2.5 mg aliquots) intravenous diazepam (0.1 mg/kg in children) or chlorpromazine 50–100 mg intramuscularly.

Amanita muscaria (fly agaric) and *Amanita pantherina* (panther cap)

Interval: 30 minutes–2 hours.

Signs of poisoning are due to two closely related compounds, ibotenic acid and muscimol. Both patterns of cholinergic and anticholinergic effects have been observed. The central nervous system features are probably mediated through γ-aminobutyric acid (GABA) receptors. The early presentation resembles alcohol intoxication (Box 10.10). Hallucinations can be severe.

Box 10.10 Pantherine syndrome

- Euphoria
- Ataxia
- Distorted visual perception
- Hallucinations
- Drowsiness and coma
- Convulsions

The stomach should be emptied as vomiting is not a clinical feature. Atropine is not indicated as both toxins potentiate its effects. Treatment is symptomatic and supportive.

Gyromitra esculenta (turban or brain fungus)

Interval: usually 6–8 hours.

Hydrolysis of the toxic principle gyromitrin yields a competitive inhibitor (monomethylhydrazine) of the coenzyme pyridoxal phosphate. Several organ systems served by this coenzyme are affected (Box 10.11).

Pyridoxine hydroxide (vitamin B_6) 25 mg/kg should be given in patients with severe or progressive symptoms.

Box 10.11 Monomethylhydrazine poisoning

- Gastrointestinal irritation
- Erythrocyte haemolysis
- Hepatic necrosis
- Renal failure

Amanita phalloides (death cap)

Interval: 12 hours approximately.

Amatoxins (cyclic octapeptides) interfere with RNA polymerase thereby inhibiting protein synthesis. They have a strong affinity for hepatocytes and the epithelial cells of the proximal convoluted tubule of the kidney. A single mushroom may contain enough amatoxin to cause death. The toxin is not significantly affected by cooking.

The classical clinical picture has two phases (Box 10.12). After a delay, gastrointestinal irritation produces vomiting, abdominal pain, and profuse watery diarrhoea. These features subside and the patient enters a period of apparent recovery lasting 24–48 hours. A deteriorating clinical state then develops as hepatic necrosis and acute renal failure develop. Muscle twitching, convulsions, and coma may then ensue. The mortality during this phase can be as high as 25 per cent.

Box 10.12 *Phalloides* syndrome

Delayed onset (12 hours)

- **Phase 1** — Gastrointestinal irritation

Latent period of 24–48 hours

- **Phase 2** — Hepatic failure and encephalopathy
  — Acute renal failure

A radioimmunoassay is available for amatoxin but treatment should begin immediately.

The stomach should be emptied if the patient presents within 36 hours of ingestion. Renal function should be monitored and fluid and electrolyte losses replaced as appropriate. Liver function should also be monitored—the prothrombin time is probably the single most useful test.

Early coma may be due to hypoglycaemia and plasma glucose should therefore be monitored.

Regular oral activated charcoal may be given to interrupt enterohepatic circulation of the toxin.

Poisoning by plants and fungi

More specific treatment of Amanita phalloides poisoning is contentious and should be discussed with medical staff at a poisons information centre. Meticulous supportive care in patients developing acute hepatic necrosis is, however, the mainstay of treatment.

Coprinus atramentarius (ink cap)

Interval: Only following consumption of alcohol, up to 72 hours after ingestion.

The toxic principle coprine is converted to 1-aminocyclopropanol which inhibits alcohol dehydrogenase. Approximately 30 minutes to 2 hours after alcohol intake features develop similar to the alcohol–disulfiram reaction due to the accumulation of acetaldehyde (Box 10.13).

Box 10.13 Disulfiram-like effects

- Facial flushing
- Sweating
- Nausea and vomiting
- Headache
- Tachyarrhythmias
- Hypotension

Propranolol 1 mg slowly intravenously over 5 minutes should be given. Up to 6 mg may be helpful for symptomatic tachyarrhythmias.

Cortinarius speciosissimus

Interval: 2–14 days.

The toxins are cyclopeptides which, after a significant delay following consumption, produce the features given in Box 10.14.

Box 10.14 *Cortinarius* poisoning

- Nausea and vomiting
- Severe thirst

- Polyuria
- Headache
- Muscle aches

Acute renal failure can develop in more serious cases and haemodialysis may be required.

Plant dermatitis and photoxicity

Most plants are harmless when in contact with the skin but a few cause irritant, allergic, and phototoxic dermatitis.

In North America, 25–60 per cent of the population are sensitive to the poison ivy and other members of the Anacardiaceae family. No single European species causes nearly as many sensitizations but dermatitis from *Primula obconica* is quite common in northern Europe. A number of indoor plants occasionally cause dermatitis. Occupational dermatitis is common in gardeners, florists, and undertakers: *Chrysanthemum* and other Compositae, *Narcissus* (daffodils and narcissi), and tulips being the most common causes. Typically the dermatitis involves the hands, forearms, and face. Often it is acute and vesicular.

Plants containing furanocoumarins are a common cause of phototoxic reactions. The furanocoumarins increase the reactivity of the skin to ultraviolet and or visible light. The phototoxic reaction is based on a non-immunological mechanism and can be elicited in the majority of individuals on first contact if the concentration of the substance and the amount of light of the appropriate wavelength are sufficient. It is a sunburn produced by wavelengths whose energy could not provoke it unless absorbed by the photosensitizer.

In the United Kingdom, primary photosensitivity is produced by contact with giant hogweed (*Heracleum mantegazzianum*) and has also occurred in people handling vegetables of the Umbelliferae family such as parsnips, carrots, and celery. Giant hogweed produces a bullous, streaky phototoxic dermatitis.

Sources should be identified and avoided, and relief from itching and inflammation provided.

Further reading

Bresinsky, A. and Besl, H. (1990). *A colour atlas of poisonous fungi*. Wolfe Publishing, London.

Cooper, M. R. and Johnson, A. W. (1984). *Poisonous plants in Britain and their effects in animals and man*. HMSO, London.

Frohne, D. and Pfander, H. J. (1984). *A colour atlas of poisonous plants*. Wolfe Publishing, London.

Proudfoot, A. T. (1993). Features and management of specific poisons. In *Acute poisoning: Diagnosis and management*, (2nd edn), pp. 61–230. Butterworth-Heinemann, Oxford.

Vale, A. and Meredith, T. (1984). Poisonous plants and mushrooms. *Medicine*, **2**, U55–9.

Chapter 11

Venomous bites and stings

General introduction	189
Venomous snakes	189
Marine stings	194
Venomous insect stings	196
Further reading	199

Key points in venomous bites and stings

1. Venom is injected in less than 50 per cent of bites by poisonous snakes and therefore providing reassurance to the victim is an important aspect of management.
2. Arterial tourniquets should not be used in snakebites. A firm ligature or bandage can be applied above the bite to delay lymphatic spread.
3. The dose of snake antivenom does not vary with the age of the victim.
4. Adrenaline should be ready and available for injection before an antivenom infusion is started.
5. A barbed bee sting should be removed immediately as the attached gland will continue to pump in venom.
6. Severe anaphylaxis following bites and stings should be treated immediately with adrenaline.
7. Standard antitetanus prophylaxis is required following bites and stings.

General introduction

Venomous bites and stings in the United Kingdom are uncommon and often trivial, but sometimes cause severe poisoning and can occasionally be fatal. On average, four to five deaths per year are the result of bites and stings from venomous animals and insects, and most of these deaths follow anaphylactic reactions to bee and wasp stings. Some bites cause considerable morbidity, with recovery taking weeks or months, especially in adults.

Rational clinical assessment of venomous bites and stings is essential as unwarranted heroic measures can have disastrous effects.

Venomous snakes

- **Venoms Adder envenomation At the scene Hospital management Antivenom Bites by non-indigenous snakes**

Snakes which are poisonous have fangs at the front of their maxillae which enable them to inject venom secreted by the parotid glands. There are three families of poisonous snakes (Table 11.1). Elapids are found in all parts of the world except Europe. Sea snakes are common in Asian and Pacific coastal waters. Pit vipers are found especially in Asia and America. True vipers are found throughout the world except in America, Asia, and Pacific areas. The European adder (*Vipera berus*) is the only native venomous snake in the United Kingdom. However, envenomation may also occur from non-indigenous species kept in reptile collections or imported inadvertently with cargo.

Venoms

Most reptile venoms consist of a complex mixture of high molecular weight proteins (enzymes), polypeptides of variable physiological activity, and low molecular weight compounds. Although snake venom activity has traditionally

Table 11.1 • Classification and features of snakebites

Snake families	Examples	Toxin	Toxic effects
Elapidae (short fixed fangs)	Cobras Mambas Coral snakes Tiger snakes	Neurotoxin	Infrequent local reaction. Muscle weakness, fasciculation and paralysis. Dysphagia. Respiratory paralysis.
Hydrophiidae (very short fixed fangs and flat tails)	Sea snakes	Myotoxin	No local reaction. Polymyositis. Myoglobinurea. Renal failure.
Viperidae			
Crotalidae (thermosensitive pit midway between eye and nostril)	Rattlesnakes Pit vipers	Vasculotoxin	Rapid local swelling and necrosis (50 per cent). Coagulation disturbance. Hypovolaemic shock.
Viperinae (long erectile fangs)	European adder Other true vipers		

been broadly grouped (Table 11.1), multiple toxic reactions can occur simultaneously or sequentially. In addition, autopharmacological agents may be released by activation of kinin and complement systems in damaged tissue and complicate the clinical picture.

The severity of the toxic effects of snakebites will depend on the extent of envenomation and quality of venom injected. The quality will in turn depend on the age and size of the snake. Furthermore, bites occurring at night are more venomous than during the day, and in the case of hibernating snakes the venom is particularly potent just after hibernation.

Adder envenomation

Immediate pain usually follows envenomation with swelling developing within minutes. Tender regional lymphadenopathy may be associated with the local reaction.

A bite wound comprising two puncture marks about 1 cm apart may be evident on close inspection and is usually located on a limb extremity.

Systemic envenomation is usually heralded by vomiting, abdominal colic, and diarrhoea. These symptoms may be accompanied by cardiovascular collapse and loss of consciousness. Systemic poisoning may occur in the absence of local evidence of envenomation.

The limb oedema can progress over 48–72 hours and become haemorrhagic. In extreme cases the swelling can include the trunk, face, and lips. Local blistering is uncommon and the venom does not cause necrosis.

In severe poisoning, hypovolaemic shock may persist for up to 36 hours and bleeding complications, oliguria, neutrophil leucocytosis, and non-specific electrocardiographic changes are evident in these cases.

Only 14 deaths from adder-bite poisoning have been recorded in the United Kingdom in 100 years.

At the scene

Probably less than 50 per cent of bites are associated with injection of venom. Reassurance of the victim is therefore one of the most important initial aspects of management. The bite should be cleaned and covered with a dry dressing. The wound should not be incised as this will lead to further local damage. Attempts to suck the poisoning from the bite are dangerous and useless. The affected limb should be immobilized and kept dependent, if at all possible, to minimize absorption of venom. If the transfer to hospital is more than 30 minutes a firm ligature or bandage can be applied above the bite to delay lymphatic spread of the venom. *A tourniquet preventing arterial circulation to the limb should not be used.*

If the snake has been killed it should be taken to the hospital for identification. However, the snake should not be handled directly even if it is dead—decapitated head reactions can persist for up to an hour and the bite from a decapitated snake can cause severe poisoning.

All victims should be referred to hospital and observed for at least 24 hours.

Hospital management

Ligatures and bandages applied in the field should be removed. After cleaning the bite wound, the area should be left exposed without a dressing. Blisters, if present, should be left intact.

Ice packs applied to the limb can aggravate local damage and should not be used. Standard antitetanus prophylaxis should be given but prophylactic antibiotics are unnecessary.

One recommended plan of investigation and monitoring is given in Box 11.1

Box 11.1 Investigation and monitoring after adder-bites

- Hourly pulse rate and blood pressure
- Measurement of urinary output and gastrointestinal fluid losses
- Daily white cell count, plasma urea, and electrolytes
- Electrocardiogram twice daily (continuous monitoring of cardiac rhythm if hypotension is present)
- Coagulation screen if there is evidence of a bleeding tendency
- Daily circumferential measurements of the proximal and distal parts of the affected limb

---

Appropriate sedation and analgesia should be provided and if hypotension supervenes intravenous volume replacement should be commenced.

Antivenom

The principal indication for the use of Zagreb antivenom is evidence of systemic poisoning:

- Fall in blood pressure (systolic to less than 80 mm Hg or by more than 50 mm Hg from the normal or admission value) with or without signs of shock
- Spontaneous bleeding

- Pulmonary oedema or haemorrhage (shown by chest radiograph)
- ECG changes (usually of ST segment and T-wave)
- Leucocytosis (greater than 20 000/mm^3)
- Elevated cardiac enzyme levels
- Evidence of renal failure

Administration of antivenom should also be considered in adults who present with established limb swelling within 2 hours of the bite as this may reduce disability from local effects.

Two ampoules (2 × 5.4 ml) of antivenom should be added to 100 ml of normal saline and initially administered at an infusion rate of 15 drops per minute. Adrenaline must be *immediately* available in a syringe beside the patient before antivenom is started. If no adverse reaction occurs the rate can be gradually increased in order to complete the infusion in approximately one hour. The dose **does not** vary with age.

Severe and immediate hypersensitivity reactions to the serum occur in about 1 per cent of patients. If a reaction occurs the infusion should be temporarily stopped and 0.3–0.5 ml of 1:1000 adrenaline (0.01 ml/kg in children) given intramuscularly, repeated if necessary. Once the symptoms of the reaction have subsided the antivenom infusion can be cautiously re-started. The dose of two ampoules can be repeated at the end of one hour if there has been no clinical improvement.

Testing for serum hypersensitivity by subcutaneous injection of a small dose of antivenom gives misleading results and is not recommended.

A history of asthma or other allergic condition is a relative contraindication to the use of antivenom and benefits have to be weighed against the risk of hypersensitivity. In these circumstances get advice from the poisons information centre, see Appendix.

Bites by non-indigenous snakes

The principles of treatment are the same as for adder-bites. Specialist advice can be obtained from poisons information centres. Stockists of antivenom against foreign snakes are given in the Appendix.

Marine stings

- **Fish stings**   Jellyfish stings

Fish stings

Several species of fish which are found in British coastal waters have venomous spines in their fins including the lesser weever fish (*Echiichthys vipera*) and sting-ray (*Dasyatis pastinaca*).

The venoms are known to contain multiple toxic fractions including serotonin.

The foot is usually affected in bathers, and fishermen may be stung if they handle these species. The immediate result of a sting is intense local pain which may spread to involve the whole limb. A puncture wound with surrounding erythema may be found. Oedema develops in a few hours and blistering and necrosis can occur. Systemic symptoms are unusual but may include vomiting, diarrhoea, hypotension, tachycardia, arrhythmias, convulsions, and respiratory depression.

The toxins are thermolabile and local symptoms can be rapidly relieved by immersing the affected part in hot water, as hot as the patient can stand for 30 minutes. Occasionally, the pain is difficult to control and titrated opiate analgesia may be required. Further relief can be obtained by infiltrating the area with 0.25 per cent bupivacaine, although this procedure may not be tolerated until systemic analgesia has been given.

Jellyfish stings

The majority of jellyfish around the coastal waters of the United Kingdom are harmless as their stings cannot penetrate human skin. Those which sting include:

- *Cyanea lamarckii* (sea nettle—blue)—lobulated bell and tufts of long tentacles
- *Cyanea capillata* (lion's mane—yellow)
- *Chrysaora hyoscella* (compass jellyfish)—Large jellyfish with radiating bands of white and brown from the centre of the bell and 24 long tentacles

The Portuguese man-of-war (*Physalia physalis*) is a hydrozoan and not a true jellyfish, which occasionally enters British waters. It comprises a blue and iridescent floating sail-shape crest (pneumatophae) which is gas-filled and from which are suspended multiple tentacles up to 30 metres in length.

The stinging organelles of jellyfish and other members of the coelenterate phylum are called nematocysts. They are located on the outer surface of tentacles (or near the mouth) and are triggered by physical contact or by chemoreceptor mechanisms.

The nematocysts contain a hollow sharply pointed and coiled tube which contains the venom. The nematocyst is contained within an outer capsule to which is attached a single pointed 'trigger'. Contact with a large Portuguese man-of-war can trigger the release of several million stinging cells. The venom contains numerous components including bradykinin, serotonin, histamine, prostaglandins, and proteases.

Following skin contact, an immediate stinging sensation occurs followed by pruritus, paraesthesia, and a burning throbbing pain which may radiate centrally from the extremities to the groin, abdomen, or axilla. The area involved by nematocysts will become red or purple often in a linear whip-like fashion corresponding to 'tentacle prints'. Other features include wheal formation, petechial haemorrhages, local oedema, and blistering. Progression to local necrosis, skin ulceration, and secondary infection can occur. Systemic symptoms may be immediate or delayed by several hours and occur as a result of direct toxicity or anaphylaxis.

The wound should be immediately rinsed with sea water but not fresh water. The area should not be rubbed with a towel or clothing to remove adherent tentacles. Both fresh water and abrasion will stimulate any nematocysts that have not already fired.

Adherent tentacles can be removed carefully with forceps or a gloved hand. Vinegar (acetic acid 5 per cent) is the treatment of choice to inactivate the toxin. This should be applied for 30 minutes or until there is no further pain. Alcohol application may worsen the situation by stimulating the discharge of nematocysts. Bathing the wound in hot water has been recommended to inactivate the thermolabile toxins, but the hypotonic solution may also cause nematocysts to fire.

No attempt should be made to restrict the venom movement by tourniquet, pressure, or immobilization. The venoms commonly cause severe pain from local injury rather than significant general effects and the venom should be allowed to disperse from the injection site.

Local anaesthetic ointments or sprays may be soothing. The patient should receive standard antitetanus prophylaxis. There is no need for prophylactic antibiotics. Standard supportive management is required if systemic complications arise. The management of anaphylaxis is outlined in Fig. 11.1.

Venomous insect stings

- **Honey bee (*Apis mellifera*)**   **Wasp (*Vespula vulgaris*)**
  Treatment

Bees and wasps are members of the order Hymenoptera. The lethal venom dose for an unsensitized human usually requires hundreds of stings. However, 0.5 per cent of the population are hypersensitive to bee or wasp venom and death can result from a single sting. Allergic reactions to stings are more common and more severe in people taking non-steroidal anti-inflammatory drugs.

The venom apparatus of Hymenoptera is located at the posterior end of the abdomen and consists of venom glands, a reservoir, and structures for piercing the skin and injecting venom.

The venom consists of mixtures of protein and polypeptide toxins, enzymes, and pharmacologically active low molecular compounds such as histamine and serotonin. Major allergens present include mellitin, phospholipase, and hyaluronidase.

Stings are most commonly inflicted on the head and neck followed by the lower and upper limbs. Immediate intense burning pain usually occurs followed by a variable amount

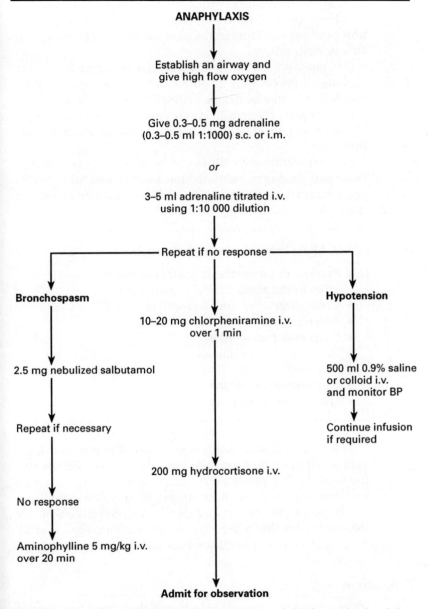

Fig. 11.1 • Immediate management of a systemic anaphylactic reaction in adults.

of local inflammation causing swelling and erythema which is seldom serious. However, occasionally accidental ingestion can lead to a sting in the pharynx causing swelling and airway obstruction.

Infrequently, the local reaction can be larger and more prolonged representing a cell-mediated hypersensitivity response. This may be incapacitating for several days and can cause dramatic changes in appearance if the face is affected. If sufficiently severe a short course of oral steroids may be indicated.

General reactions to stings can be more serious. Some of these may be due to vasovagal reactions to painful or frightening stings, but severe anaphylactic reactions do occur and must be recognized early (Box 11.2).

Box 11.2 Clinical features of anaphylaxis

- Pruritus or paraesthesia (distant to the sting and often in the scalp, palate, or perineum)
- Generalized erythema and urticaria
- Angiooedema
- Dyspnoea and wheeze
- Stridor and vocal changes
- Diarrhoea and vomiting
- Cardiovascular collapse
- Loss of consciousness

Systemic reactions may occur in the absence of a local skin reaction to the sting. Life-threatening reactions are less frequent in children than in adults. Death from a systemic reaction following a hymenopterous sting is due to multiple mechanisms and not simply due to anaphylaxis alone. Of those who die after a bee sting, 90 per cent are aged over 25 years and atherosclerosis appears to be a major associated factor.

Treatment

If the sting is still present (a feature of the barbed bee sting) it should be removed, as the attached gland will still continue to pump venom. Removal can be achieved by flicking the sting with a knife blade or a fingernail edge rather than by grasping it with fingers or forceps as this may squeeze in more venom.

Prompt treatment of anaphylaxis should be started according to the plan outlined in Fig. 11.1.

Bee venom immunotherapy may be offered to survivors of a systemic reaction with demonstrable IgE antibodies or in those cases where there is increasing severity of reaction to successive stings. Immunotherapy has, however, not been proven to prevent death from a bee sting. Some patients opt to carry a preloaded adrenaline syringe or adrenaline inhaler instead.

Further reading

Cain, D. (1983). Weever fish sting; an unusual problem. *British Medical Journal*, **287**, 406–7.

Fisher, M. (1992). Treating anaphylaxis with sympathomimetic drugs. *British Medical Journal*, **305**, 1107–8.

Proudfoot, A. (1993). Adder envenomation. In *Acute poisoning: Diagnosis and management*, pp. 62–4. Butterworth-Heinemann, Oxford.

Reid, H. A. (1976). Adder bites in Britain. *British Medical Journal*, **2**, 153–6.

Rubenstein, H. S. (1982). Bee sting diseases: Who is at risk? What is the treatment? *Lancet*, **1**, 496–9.

Sutherland, S. K. (1984). Management of venomous bites and stings. *Medicine International*, **2**, 391–6.

Warrell, D. A. (1979). Bites and stings by venomous animals in Britain. *Prescribers Journal*, **19**(6), 190–9.

Warrell, D. A. (1987). Injuries, envenoming, poisoning and allergic reactions caused by animals. In *Oxford textbook of medicine* (ed. D. J. Weatherall, J. G. G. Ledingham, and D. A. Warrell), (2nd edn), pp. 6.66–85. Oxford University Press.

Youlten, L. J. F. (1987). Anaphylaxis to bee and wasp stings. *The Practitioner*, **231**, 502–4.

Appendix:
Sources of help

Poisons information centres

Eire
Dublin
Poisons Information Centre Tel. 010 353 1 379966
Beaumont Hospital Fax 010 353 1 438 346
PO Box 1297
Beamont Road
Dublin 9

England
Birmingham
West Midlands Poisons Unit Tel. 021 554 3801
Dudley Road Hospital Fax 021 507 5580
Birmingham B18 7QH

Leeds
The Leeds Poisons Information Tel. 0532 430715
   Service Fax 0532 445849
Pharmacy Department
The General Infirmary
Gt. George Street
Leeds LS1 3EX

London
National Poisons Information Tel. 071 635 9191
   Service Fax 071 635 1053
London Centre
Avonley Road
London SE14 5ER

Newcastle
Northern Region Drug and Tel. 091 232 5131
   Therapeutic Centre Fax 091 261 5733
Royal Victoria Infirmary
Queen Victoria Road
Newcastle-Upon-Tyne NE1 4LP

Northern Ireland
Belfast
Poisons Information Centre Tel. 0232 240503
Royal Victoria Infirmary Fax 0232 248030
Grosvenor Road
Belfast BT12 6BA

Scotland
Edinburgh
Scottish Poisons Information Bureau Tel. 031 229 2477
The Royal Infirmary Fax 031 228 3332
Lauriston Place
Edinburgh EH3 9YW

Wales
Cardiff
Welsh National Poisons Unit Tel. 0222 709901
Ward West 5 Fax 0222 704357
Llandough Hospital
Penarth
South Glamorgan CF6 1XX

Toxbase

Toxbase is a computerized information guide to the ingredients, toxicity, features, and treatment of intoxication with a variety of products including drugs, chemicals, pesticides, household products, and plants. The database contains information on about 8000 products and is constantly updated by clinical toxicologists working in the UK poisons information services. The information is stored on a computer in Edinburgh and is a *viewdata* system. Viewdata is the technology whereby information can be accessed using a terminal linked to the computer by the national telephone network or through the ISTEL Infotrac network which allows users access to the computer at the cost of local telephone calls.

Further information can be obtained from:

Scottish Poisons Information Bureau
The Royal Infirmary
Lauriston Place
Edinburgh EH3 9YW

Foreign snake antivenom

- For information on identification, management, and supply, telephone:

    National Poisons Information Service, London Centre, tel. 071 635 9191

    Pharmacy
    Walton Hospital
    Rice Lane
    Liverpool L9 1AE
    (supply only)

    Tel. 051 525 3611
    Fax 051 529 4445

    Liverpool School of Tropical Medicine
    Penbroke Lane
    Liverpool L35 QA

    Tel. 051 708 9393
    Fax 051 708 8733

    Centre for Tropical Medicine
    John Radcliffe Hospital
    Headington
    Oxford OX3 9DU

    Tel. 0865 220968
    Fax 0865 220984

- Antivenom for adder-bites is held widely and advice can be obtained from poisons information centres.

Information on diving emergencies and recompression chambers

- HM Coastguard (tel. 999) will provide a useful communication link for a doctor in a remote location faced with a seriously ill diver.
- For use of naval, commercial, and experimental hyperbaric chambers in the United Kingdom and advice on regimen and management contact:

    Duty Diving Medical Officer
    HMS *Vernon*
    Portsmouth PO1 3HH

    Tel. 0705 818888 (emergency) or 0705 822351 (routine enquiries) extension 41584 during work hours

- Other sources of assistance

Diving Emergency Services
Diving Diseases Research Centre
Fort Bovisand
Plymouth PL9 0AB

Tel. 0752 261910 (Aircall 24 hour emergency number). Ask for the Duty Diving Doctor.

Hyperbaric Medicine Unit
Aberdeen Royal Hospitals NHS Trust
Foresterhill
Aberdeen AB9 2ZB

Tel. 0224 681818
State that you have a diving emergency and give your name and telephone number. Contact will then be made with the Duty Hyperbaric Physician.

Regional Recompression Unit
Craigavan Hospital
Co. Armagh BT63 5QQ
Northern Ireland

Tel. 0762 336711 (direct line)

UIAA Mountain Medicine Centre

The UIAA (International Union of Alpine Associations) Mountain Medicine Centre offers information to mountaineers, trekkers, and mountaineering doctors on the following:

- Acclimatization and acute mountain sickness
- High-altitude pulmonary and cerebral oedema
- Frostbite and hypothermia
- Expedition medical equipment
- Injuries and deaths during mountaineering
- Problems of extreme altitude
- Medical problems of trekking and mountain holidays
- Fitness to travel to high altitudes

Contact:
Dr Charles Clarke, FRCP,
UIAA Mountain Medicine Centre
St. Bartholomew's Hospital
London EC1A 7BE

Index

acclimatization 75, 91, 93
acetazolamide (in high altitude illness) 90, 99–100
acids, see corrosives
acrolein 132
acute altitude illness 96–101
 prevention 99–101
 treatment 98–9
 types 96–7
acute mountain sickness 96
acute radiation syndrome 143, 148–51
adder, European (Vipera berus) 189, 190–1
adder-bite, see envenomation
adult respiratory distress syndrome 41, 49
air embolism 60
airway warming 16–17
albedo 101
alkalis, see corrosives
alkaloids 171–4, 178–9
Amanita muscaria (fly agaric) 182
Amanita pantherina (panther cap) 182
Amanita phalloides (death cap) 179, 183–4
amatoxin 183
ammonia 126
anaphylaxis (after insect stings)
 clinical features 198
 management 197
anhidrotic heat exhaustion 77
antimuscarinic syndrome 174
antivenom 192–3
 hypersensitivity to 193
Araceae spp. 179
arsenical vesicants 136

barotrauma 55, 57–60
 of ascent 59–60
 of descent 57–9
bee, see honey bee
'bends', the 61

bites, see venomous bites and stings
black bryony (Tamus communis) 179
black nightshade (Solanum nigrum) 178
bleeding diathesis 81
brain fungus, see Gyromitra esculenta
British Anti-Lewisite (BAL), see dimercaprol
broom (Cystisus scoparius) 173

carbon monoxide poisoning 127–30
 hyperbaric oxygen therapy in 128–30
carbamates, see insecticides
cardiac glycosides 174–5
cardiopulmonary bypass 18, 85
charcoal, activated 169, 183
CHEMDATA 118
CHEMET 118
chemical accidents and emergencies 115–38
 decontamination 120
 hazard identification 118
 resuscitation 120–1
 safety protocols 118–20
chemical contamination 121–6
 chemical burns of the eye 125–6
 corrosives 121–3
 mercurials 124–5
 organophosphate and carbamate insecticides 123–4
chemical weapons, injury from 135–8
 nerve agents 136–7
 riot control agents 137–9
 vesicant agents 135–6
chilblain (perniosis) 32
chlorine 126
'chokes', the, see decompression sickness
cold injury 28–37
 determining factors in 28–30
 freezing vs. non-freezing 28
 types of 28

cold stress 3, 93
contaminated (irradiated) casualty, the 152–9
  arrangements for decontamination in hospital 153–5
  decontamination 155–9
  field management 152–3
Control of Industrial Major Accident Hazard Regulations 1984 (CIMAH) 117
cooling, methods of 83–4
core temperature
  'afterdrop' in hypothermia 11
  measurement of 22
  regulation of 5–7
cotoneaster (*Cotoneaster spp.*) 176
coumarine glycosides 179
countermeasures (after environmental release of radiation) 160–2
  emergency reference levels (ERLs) and, 160–1
cowbane (*Cicuta virosa*) 177
CS (tear) gas 137
cyanide poisoning 130–1, 176
  antidote 131, 177
cyanogenic glycosides 176–7

deadly nightshade (*Atropa belladonna*) 174
death cap, *see Amanita phalloides*
decompression sickness 60–7
  central nervous system, pulmonary, and vestibular 62–3
  cutaneous 62
  joint ('the bends') 61–2
  treatment 63–7
decontamination
  after chemical accidents and emergencies 120
  after radiation accidents 155–9
dicobalt edetate 131, 177
dimercaprol (British Anti-Lewisite, BAL) 125, 136
diving emergencies 51–67
  field management 63–4
  hospital management 65–7
diving physics 54–6
  gas laws and 54–6
  partial pressures 56
drowning 41
  dry drowning 41
  near drowning 40–50
  secondary drowning 41, 49
  see also adult respiratory distress syndrome
dysbarism 53

elderberry (*Sambucus spp.*) 176
emergency reference levels (ERLs), *see* countermeasures
emesis, induction of, 169
entonox; after diving accidents 64
envenomation, adder 190–1
  antivenom 192–3
  field management 191
  hospital management 192
European adder, *see* adder
extracorporeal warming 18
eyes, chemical burns 124–6
  irrigation 125

flashover effect 106, 111
fly agaric, *see Amanita muscaria*
foxglove (*Digitalis purpurea*) 174
frostbite 28, 33–6
  clinical features 34
  field management 34–5
  hospital management 35–6
  pathophysiology 33–4
frostnip 34
fungi, *see* poisonous fungi
fungicides, *see* mercurials

giant hogweed (*Heracleum mantegazzianum*) 185
glycoalkaloids 178–9
glycosides 174–5, 176–7, 179
ground current (step voltage) injury 110
Gyromitra esculenta (turban or brain fungus) 182

HAZ CHEM codes 118
heat accumulation 72
heat cramps 76
heat exhaustion 77–9
heat illness 69–87
  heat stroke 79–86
  minor heat illness 75–9
heat loss 5–7, 72–4
heat oedema 75–6

Index • 209

heat physiology 72–4
  see also thermoregulation
heat production (thermogenesis) 6–7, 72
heat stress, measurement 86
heat stroke 79–86
  classical 79
  clinical features 80–2
  differential diagnosis 80
  exertional 79
  management 82–5
  prevention 85–6
heat syncope 77
heat tetany 76–7
hemlock (*Conium maculatum*) 172
hemlock water dropwort (*Oenanthe crocata*) 177
high-altitude cerebral oedema 97
high-altitude illness 89–104
high-altitude pulmonary oedema 96–7
honey bee (*Apis mellifera*) 196
horse chestnut (*Aesculus hippocastanum*) 179
hydrofluoric acid 122
hydrogen chloride 126
hydrogen fluoride 126
hyperkalaemia, in cardiac glycoside poisoning 175
hypocapnia 94
hypothalamus 6, 74
hypothermia 2–25
  clinical features 7–10
  differentiation from death 10–11, 19
  field management 19–21
  hospital management 22–4
  laboratory findings 10–11, 23
  rewarming techniques 13–18
  types 3–4
hypoxia (high altitude), response to, 93–5

immersion accidents 41–2
  predisposing factors 42
  terminology 41
immersion foot, see trench foot
immunotherapy, bee venom 199
insecticides 123–4
intermittent positive pressure ventilation (IPPV) 47
iodine (radioactive), prophylaxis for, see stable iodine
ionizing radiation 141–5, 147–8
  biological effects 147–8
  ionizing density 143–4
  quantities and units 144–5
  types 142–3
ipecacuanha, syrup of, 169
irrigation, ocular 125

laburnum (*Laburnum anagyroides*) 173
Lichtenberg figures 106, 112
lightning conductor rod 110
lightning discharge 108–9
lightning injuries 105–14
  avoidance 113–14
  diagnosis 111–13
  incidence 107
  management 113
  mechanism of injury 110–11
  predisposing factors 107
lily of the valley (*Convallaria majalis*) 174
lime (calcium oxide) 122
lupin (*Lupinus spp.*) 173

'magic mushrooms' 181
mediastinal emphysema 59
mercurials 124–5
  antidotes 125
mistletoe (*Viscum album*) 175
monkshood (*Aconitum napellus*) 178
monomethylhydrazine poisoning 182
mountain sickness, see acute mountain sickness
muscarinic effects (after mushroom poisoning) 181
mustard gas 135–6

near drowning 40–50
  complications 49–50
  hospital investigations 46
  hypothermia and 47
  prognosis 50
  pulmonary injury 42–3
  resuscitation 44–8
  sea v. fresh water aspiration 43
nematocyst 195
nerve agents 136–7
nicotinic effects (after plant poisoning) 172
nitrogen dioxide 126
nitrogen mustard, see mustard gas
nitrogen narcosis 56, 58–9
nuclear weapons accident 162

occupational (plant) dermatitis 185
oleander (*Nerium oleander*) 175
organophosphate and carbamate insecticides 123–4
  antibodies 124
organophosphates, *see* insecticides
oxalates 179

pantherine syndrome 182
penicillamine 125
perilymphatic fistula 57, 58
periodic breathing 95
peritoneal dialysis, rewarming by, 17, 85
phalloides syndrome 183
phenol (carbolic acid) 122–3
phosgene (carbonyl chloride) 126
phototoxicity 185
plutonium 162
pneumothorax 60
poisoning by plants and fungi 165–86
  fungi 179–85
  general approach to management 167–70
  incidence 167
  plant dermatitis and phototoxity 185
  plants that cause antimuscarinic effects 174
  plants that cause cardiovascular disturbance 174–6
  plants that cause CNS disturbance 177–8
  plants that cause cyanide poisoning 176–7
  plants that cause gastrointestinal irritation 178–9
  plants with nicotinic effects 172–3
poisonous fungi 179–85
  Amanita spp. 182–3
  Coprinus atramentarius 184
  Cortinarius speciosissimus 184–5
  gastrointestinal irritants 180
  Gyromitra esculenta 182
  Inocybe and *Clitocybe* spp. 180–1
  Psilocybe and *Panaeolus* spp. 181
portuguese man-of-war (*Physalia physalis*) 195
positive end expiratory pressure (PEEP) ventilation 49
potatoes (*Solanum tuberosum*) 178
pralidoxime 124

pressurization bag 98–9
prickly heat 77
protective clothing
  for chemical emergencies 120
  for radiation accidents 154
Prunus spp. 176
pulmonary barotrauma 59
pyridine–piperidine alkaloids 172
pyridostigmine 137

quicklime 122
quinolizidine alkaloids 173

radiation accidents 139–63
  contingency plans 155
  environmental release ('off-site' radiation exposure) 159–62
  irradiated v. contaminated casualties 146, 151–9
  nuclear weapons 162
  sources 141
  types 145–6
recompression 63, 66–7
renin-angiotensin-aldosterone system 75, 95
resins 177–8
respiratory alkalosis 94, 100
retinal haemorrhages 97, 127
rewarming techniques
  in hypothermia 13–18
  in frostbite 35–6
rhabdomyolysis 81

secondary drowning, *see* adult respiratory distress syndrome
self-contained underwater breathing apparatus (SCUBA) 53–4, 55–6
shivering 7
side-flash 110
smoke
  inhalation 131–3
  combustion, materials, and products of 132
snake bites
  bites by non-indigenous snakes 193
  classification and features 189–90
  see also envenomation
snowblindness 102–3
solvent abuse ('glue sniffing'), *see* volatile hydrocarbons

spurge laurel (*Daphne laureola*) 179
spurge olive (*Daphne mezereum*) 179
'squeeze', the, 57
   see barotrauma of descent
'staggers', the
   see decompression sickness
stings
   bee 196–9
   fish 194
   jellyfish 194–6
   wasp 196–9
sulphur dioxide 126
sulphur mustard, see mustard gas
sun protection factor (SPF) 102
sunburn 101–2
sunscreens 102
swiss cheese plant (*Monstera*) 179

tear gas 137
thermoregulation 5–7, 72–4
thornapple (*Datura stramonium*) 174
thunder 110
toxic inhalations 126–36
   carbon monoxide 127–30
   cyanide 130–1
   hydrogen sulphide 133
   respiratory irritants 126–7
   smoke 131–3
   volatile hydrocarbons 133–5
transport emergency (TREM) cards 118
trench foot (immersion foot) 30–2
Trendelenberg position, the 63
tropane alkaloids 174

ultraviolet radiation
   injury from 101–3
   types 101

venomous bites and stings 187–99
   incidence 189
   insect stings 196–9
   marine stings 194–6
   venomous snakes 189–93
vertigo, post-dive
   causes of 62–3
viscotoxins 175–6
volatile hydrocarbons 133–5

wasp (*Vespula vulgaris*) 196
water hemlock, see cowbane
wet bulb globe temperature (WBGT) index 86
wind chill 5, 29–31, 93
woody nightshade (*Solanum dulcamara*) 178

yew (*Taxus baccata*) 178

Zagreb antivenom 192